T0201234

GUIDELINES FOR INTEGRATING MANAGEMENT SYSTEMS AND METRICS TO IMPROVE PROCESS SAFETY PERFORMANCE

This book is one in a series of process safety guidelines and concept books published by the Center for Chemical Process Safety (CCPS). Please go to www.wiley.com/go/ccps for a full list of titles in this series.

GUIDELINES FOR INTEGRATING MANAGEMENT SYSTEMS AND METRICS TO IMPROVE PROCESS SAFETY PERFORMANCE

Center for Chemical Process Safety
New York, NY

WILEY

Published by John Wiley & Sons, Inc., Hoboken, New Jersey.
Published simultaneously in Canada.

For general information on our other products and services please contact our Customer Care Department within the United States at (800) 762-2974, outside the United States at (317) 572-3993 or fax (317) 572-4002.

Wiley also publishes its books in a variety of electronic formats. Some content that appears in print, however, may not be available in electronic formats. For more information about Wiley products, visit our web site at www.wiley.com.

Library of Congress Cataloging-in-Publication Data:

American Institute of Chemical Engineers. Center for Chemical Process Safety, author.
  Guidelines for integrating management systems and metrics to improve process safety performance / Center for Chemical Process Safety of the American Institute of Chemical Engineers.
      pages cm
  Includes index.
  ISBN 978-1-118-79503-3 (cloth)
  1. Industrial safety—Management. 2. Chemical processes—Safety measures. 3. Systems integration. I. Title.
  T55.A454 2015
  658.3'82—dc23   2015012596

Printed in the United States of America.

10 9 8 7 6 5 4 3 2

CONTENTS

LIST OF FIGURES

LIST OF TABLES

ACRONYMS AND ABBREVIATIONS

ACC	American Chemistry Council
AFPM	American Fuel & Petroleum Manufacturers
AIChE	American Institute of Chemical Engineers
ALARP	As Low As Reasonably Practicable
API	American Petroleum Institute
BPCS	Basic Process Control System
CCPS	Center for Chemical Process Safety
CFR	Code of Federal Regulations
COMAH	UK HSE Control of Major Accident Hazards
CPI	Chemical Process Industries
CSB	U.S. Chemical Safety Board
EPA	U.S. Environmental Protection Agency
EU	European Union
ISO	International Organization for Standardization
OD	Operational Discipline
OECD	Organisation for Economic Co-operation and Development
OSHA	U.S. Occupational Safety and Health Administration
PSM	Process Safety Management
RAGAGEP	Recognized and Generally Accepted Good Engineering Practices
RBPS	Risk Based Process Safety
RC	Responsible Care®
RMP	Risk Management Program
SHEQ&S	Safety (process safety), Health (occupational safety and health), Environmental, Quality, and Security
UK	United Kingdom
UK HSE	UK Health and Safety Executive
U.S.	United States

GLOSSARY

This Glossary is current at the time of publication. Please access the CCPS website for the latest Glossary.

Accident Prevention Pillar	A group of mutually supporting RBPS elements. The RBPS management system is composed of four accident prevention pillars: (1) commit to process safety, (2) understand hazards and risk, (3) manage risk, and (4) learn from experience.
Administrative Control	Procedures that will hold human and/or equipment performance within established limits.
Barrier	Anything used to control, prevent, or impede energy flows. Includes engineering (physical, equipment design) and administrative (procedures and work processes). See also Layer of Protection.
Basic Process Control System (BPCS)	A system that responds to input signals from the process and its associated equipment, other programmable systems, and/or from an operator, and generates output signals causing the process and its associated equipment to operate in the desired manner and within normal production limits.
Bow Tie	A diagram for visualizing the types of preventive and mitigative barriers which can be used to manage risk. These barriers are drawn with the threats on the left, the unwanted event at the center, and the consequences on the right, representing the flow of the hazardous materials or energies through its barriers to its destination. The hazards or threats can be proactively addressed on the left with specific barriers (safeguards, layers of protection) to help prevent a hazardous event from occurring; barriers reacting to the event to help reduce the event's consequences are shown on the right.

Consequence	The direct, undesirable result of an accident sequence usually involving a fire, explosion, or release of toxic material. Consequence descriptions may be qualitative or quantitative estimates of the effects of an accident.
Consequence Analysis	The analysis of the expected effects of incident outcome cases, independent of frequency or probability.
Containment	A system condition in which under no condition reactants or products are exchanged between the chemical system and its environment.
Engineered Control	A specific hardware or software system designed to maintain a process within safe operating limits, to safely shut it down in the event of a process upset, or to reduce human exposure to the effects of an upset.
Environmental Group	In context of this guideline, the environmental group manages the air, water and land permits, including hazardous waste storage and disposal.
Equipment	A piece of hardware which can be defined in terms of mechanical, electrical or instrumentation components contained within its boundaries.
Equipment Reliability	The probability that, when operating under stated environment conditions, process equipment will perform its intended function adequately for a specified exposure period.
Event	An occurrence involving a process that is caused by equipment performance or human action or by an occurrence external to the process.
Explosion	A release of energy that causes a pressure discontinuity or blast wave.
Explosive	A chemical that causes a sudden, almost instantaneous release of pressure, gas, and heat when subjected to sudden shock, pressure, or high temperature. (OSHA 1994)
Facility	The physical location where the management system activity is performed. In early life-cycle stages, a facility may be the company's central research laboratory or the engineering offices of a technology vendor. In later stages, the facility may be a typical chemical plant, storage terminal, distribution center, or corporate office. Site is used synonymously with facility when describing RMP audit criteria.
Failure	An unacceptable difference between expected and observed performance.
Fire	A combustion reaction accompanied by the evolution of heat, light and flame.

Fire Protection	Methods of providing for fire control or fire extinguishment. (NFPA 850)
Flammable	A gas that can burn with a flame if mixed with a gaseous oxidizer such as air or chlorine and then ignited. The term "flammable gas" includes vapors from flammable or combustible liquids above their flash points.
Frequency	Number of occurrences of an event per unit time (e.g., 1 event in 1000 yrs. $= 1 \times 10^{-3}$ events/yr.).
Hazard	An inherent chemical or physical characteristic that has the potential for causing damage to people, property, or the environment. In this document it is the combination of a hazardous material, an operating environment, and certain unplanned events that could result in an accident.
Hazard Analysis	The identification of undesired events that lead to the materialization of a hazard, the analysis of the mechanisms by which these undesired events could occur, and usually the estimation of the consequences.
Hazard Evaluation	Identification of individual hazards of a system, determination of the mechanisms by which they could give rise to undesired events, and evaluation of the consequences of these events on health (including public health), environment, and property. Uses qualitative techniques to pinpoint weaknesses in the design and operation of facilities that could lead to incidents.
Hazardous Material	In a broad sense, any substance or mixture of substances having properties capable of producing adverse effects to the health or safety of human beings or the environment. Material presenting dangers beyond the fire problems relating to flash point and boiling point. These dangers may arise from, but are not limited to, toxicity, reactivity, instability, or corrosivity.
Health Group	In context of this guideline, the group administering the Occupational Safety and Health programs.
Incident	An event, or series of events, resulting in one or more undesirable consequences, such as harm to people, damage to the environment, or asset/business losses. Such events include fires, explosions, releases of toxic or otherwise harmful substances, and so forth.
Independent Protection Layer (IPL)	A device, system, or action that is capable of preventing a postulated accident sequence from proceeding to a defined, undesirable endpoint. An IPL is independent of the event that initiated the accident sequence and independent of any other IPLs. IPLs are normally identified during layer of protection analyses.

Layer of Protection	A device, system, or action, supported by a management system that is capable of preventing an initiating event from propagating to a specific loss event or impact.
Layer of Protection Analysis (LOPA)	An approach that analyzes one incident scenario (cause-consequence pair) at a time, using predefined values for the initiating event frequency, independent protection layer failure probabilities, and consequence severity, in order to compare a scenario risk estimate to risk criteria for determining where additional risk reduction or more detailed analysis is needed. Scenarios are identified elsewhere, typically using a scenario-based hazard evaluation procedure such as a HAZOP Study.
Likelihood	A measure of the expected probability or frequency of occurrence of an event. This may be expressed as an event frequency (e.g., events per year), a probability of occurrence during a time interval (e.g., annual probability) or a conditional probability (e.g., probability of occurrence, given that a precursor event has occurred).
Mitigation	Lessening the risk of an accident event sequence by acting on the source in a preventive way by reducing the likelihood of occurrence of the event, or in a protective way by reducing the magnitude of the event and/or the exposure of local persons or property.
Normal Process	Any process operations intended to be performed between startup and shutdown to support continued operations within the safe upper and lower operating limits.
Occupational Safety and Health	In context of this guideline, the discipline that focuses on the prevention and mitigation of adverse health effects on people working with hazardous materials and energies, such as industrial hygiene and personal protective equipment. This discipline also addresses safe work practices, such as confined space entry, electrical energy isolation, line breaks and fall protection. (Compare to the process safety discipline).
Quality Group	In context of this guideline, the group in an organization that monitors the quality of the product, including such management systems as ISO 9000, and ensuring customer relations.
Pillar	See Accident Prevention Pillar
Prevention	The process of eliminating or preventing the hazards or risks associated with a particular activity. Prevention is sometimes used to describe actions taken in advance to reduce the likelihood of an undesired event.

Process Safety	A disciplined framework for managing the integrity of operating systems and processes handling hazardous substances by applying good design principles, engineering, and operating practices. It deals with the prevention and control of incidents that have the potential to release hazardous materials or energy. Such incidents can cause toxic effects, fire, or explosion and could ultimately result in serious injuries, property damage, lost production, and environmental impact.
Process Safety System (PSS)	A process safety system comprises the design, procedures, and hardware intended to operate and maintain the process safely.
Process Safety Management (PSM)	A management system that is focused on prevention of, preparedness for, mitigation of, response to, and restoration from catastrophic releases of chemicals or energy from a process associated with a facility.
Program	A series of actions proposed in order to achieve a certain result.
Reliability	The probability that an item is able to perform a required function under stated conditions for a stated period of time or for a stated demand.
Risk	A measure of human injury, environmental damage, or economic loss in terms of both the incident likelihood and the magnitude of the loss or injury. A simplified version of this relationship expresses risk as the product of the likelihood and the consequences of an incident. (i.e., Risk = Consequence × Likelihood)
Risk Based Process Safety (RBPS)	The Center for Chemical Process Safety's process safety management system approach that uses risk-based strategies and implementation tactics that are commensurate with the risk-based need for process safety activities, availability of resources, and existing process safety culture to design, correct, and improve process safety management activities.
Runaway Reaction	A thermally unstable reaction system which exhibits an uncontrolled accelerating rate of reaction leading to rapid increases in temperature and pressure.
Safeguards or Protective Features	Design features, equipment, procedures, etc. in place to decrease the probability or mitigate the severity of a cause-consequence scenario.
Safety Group	In context of this guideline, the safety group is divided between the process safety and the occupational safety and health disciplines.

Safety Layer	A system or subsystem that is considered adequate to protect against a specific hazard. The safety layer cannot be compromised by the failure of another safety layer, it is totally independent of any other protective layers, may be a non-control alternative (e.g., chemical, mechanical), may be an administrative procedure, may require diverse hardware and software packages, must be approved according to company policy and procedures, must have acceptable reliability, and must meet proper equipment classification.
Safety System	Equipment and/or procedures designed to limit or terminate an incident sequence, thus avoiding a loss event or mitigating its consequences.
Security Group	In context of this guideline, the security group manages and controls access to the facility.
Shutdown	A process by which operations are brought to a safe and non-operating condition.
System	A collection of people, equipment and methods organized to accomplish a set of specific functions.
Toller	A contracted company that manufactures, stores, uses, handles, or transports chemical components of a facility's final products. Sometimes called third party service provider, toll processor, supplier of outside services, external contract manufacturer, contract processor, contract manufacturer, custom chemical manufacturer.
Toxic Hazard	In the context of these guidelines, a measure of the danger posed to living organisms by a toxic agent, determined not only by the toxicity of the agent itself, but also by the means by which it may be introduced into the subject organisms under prevailing conditions.
Toxic Material	A material that, when exposed to living organisms at a specified dose, has the potential to cause injury or death (it is poisonous).
Unstable Material	A material that, in the pure state or as commercially produced, will vigorously polymerize, decompose or condense, become self-reactive, or otherwise undergo a violent chemical change under conditions of shock, pressure, or temperature. (NFPA 704, 2001 edition)

ACKNOWLEDGMENTS

The American Institute of Chemical Engineers (AIChE) and the Center for Chemical Process Safety (CCPS) express their appreciation and gratitude to all members of this CCPS Subcommittee for Project (P 247) and their CCPS member companies for their generous support and technical contributions in the preparation of these Guidelines. The AIChE and CCPS also express their gratitude to the team of authors from BakerRisk.

Subcommittee Members:

Tony Downes	Committee Co-Chair, Honeywell
Jeff Fox	Committee Co-Chair, Dow Corning
Habib Amin	Costa Contra County
Steve Arendt	ABS Consulting
Edward Dyke	Merck
Wayne Garland	Eastman
David Guss	Nexen
David Moore	Acu Tech Consulting
Cathy Pincus	Exxon Mobil
Patricia Shaw	Koch Industries
Della Wong	CNRL
Dave Belonger	Staff Consultant, CCPS

CCPS wishes to acknowledge the many contributions of the BakerRisk staff members who wrote this book, especially the principal author Bruce K. Vaughen and his colleagues Thomas Rodante, David Black, Michael Broadribb, Michael Johnston, Charlie Pacella and Jatin Shah. Editing assistance from Moira Woodhouse, BakerRisk, is gratefully acknowledged, as well.

Before publication, all CCPS books are subjected to a thorough peer review process. CCPS gratefully acknowledges the thoughtful comments and suggestions of the peer reviewers. Their work enhanced the accuracy and clarity of these guidelines.

Peer Reviewers:

Christopher Conlon	National Grid
Jonas Duarte	Chemtuira Corporation
Bob Gregorovich	Wesfarmers, Chemicals, Energy & Fertiliser
Dan Miller	BASF Corporation, CCPS TSC member
Keith R Pace	Praxair
Richard E. Stutzki	3M
Michael Vopatek	LyondellBasell
Toni Wenzel	Honeywell International

PREFACE

The American Institute of Chemical Engineers (AIChE) has been closely involved with process safety and loss control issues in the chemical and allied industries for more than four decades. Through its strong ties with process designers, constructors, operators, safety professionals, and members of academia, AIChE has enhanced communications and fostered continuous improvement of the industry's high safety standards. AIChE publications and symposia have become information resources for those devoted to process safety and environmental protection.

AIChE created the Center for Chemical Process Safety (CCPS) in 1985 after the chemical disasters in Mexico City, Mexico, and Bhopal, India. The CCPS is chartered to develop and disseminate technical information for use in the prevention of major chemical accidents. The center is supported by more than 150 chemical process industries (CPI) sponsors who provide the necessary funding and professional guidance to its technical committees. The major product of CCPS activities has been a series of guidelines to assist those implementing various elements of a process safety and risk management system. This book is part of that series.

The CCPS Technical Steering Subcommittee overseeing this guideline was chartered to review and update the 1996 CCPS book, Guidelines for Integration of PSM, ES&H and Quality. This guideline has been written to reflect the increased attention for security at facilities handling hazardous materials and to capture the recent advances in understanding how process safety performance improvements can be measured with a combination of leading and lagging indicators. Since the management programs for the process safety, occupational safety and health, environmental, quality and security groups have developed separately in many organizations, this guideline has been written to help organizations identify metrics which affect process safety performance across the SHEQ&S groups. Integrating these metrics will reduce an organization's overall operational risks.

You can access tools, templates and documents for *Guidelines for Integrating Management Systems and Metrics to Improve Process Safety Performance* at the CCPS Website:

http://www.aiche.org/ccps/publications/metrics-tools

1 INTRODUCTION

Since its founding in 1985, the Center for Chemical Process Safety (CCPS) of the American Institute of Chemical Engineers (AIChE) has promoted the enhanced management of chemical process safety. The CCPS has always recognized that good safety performance is achieved through a combination of technology and management excellence.

The management programs for the process safety, occupational safety and health, environmental, quality and security groups have developed separately in many organizations. CCPS recognizes that significant overall operational risk reduction occurs when these programs establish common management systems and metrics across the groups managing them. Hence, merging the similarities and common needs of these different programs will lead to more efficient and effective management within the organization. This guideline provides both small and large organizations with approaches to help identify and evaluate and leverage the common systems and metrics across the groups based on the hazards and risks being monitored for each group.

1.1 THE NEED FOR INTEGRATION

Many companies have overlapping regulatory, industry and trade association, and certification requirements that can consume significant resources and attention. Identifying synergies between these performance improvement systems will help ensure safe and reliable operations, will help streamline procedures and cross-system auditing, and will support regulatory and corporate compliance requirements. Since some of the systems and metrics are common to more than one function, a well-designed and implemented integrated management system will help reduce the load on the process safety, occupational safety and health, environmental, quality and security groups. In addition, an integrated system will help improve manufacturing efficiency and customer satisfaction. Integration of process safety, occupational safety and health, environmental, quality and security performance improvement systems have been noted in recent metrics-related themes at conferences, webinars, journals and books.

In almost every region and industrialized country, regulations have been introduced that require formal process safety, occupational safety and health, environmental and security management programs. Examples for process safety regulations include: the U.S. OSHA Process Safety Management (PSM) Standard and U.S. EPA Risk Management Program (RMP), the Canadian EPA Environmental Emergency Regulations, and the European Directive Seveso II. Detailed reference lists, included in Appendix A, provide a summary of U.S. regulations (Table A-1), international regulations (Table A-2), voluntary industry standards (Table A-3), consensus codes (Table A-4) and organizations committing efforts to process safety (Table A-5).

Whether a facility is regulated or not, if it must handle hazardous materials and energies, a company's success will be impacted by how well it applies the fundamental elements of a process safety and risk management system and integrates metrics which affect process safety performance with its other risk reduction programs. As is shown in Table 1-1, the "business case" for process safety has been noted by several organizations (ACC 2013a, CCPS 2006) and was succinctly stated by Trevor Kletz decades ago, with many variations since then: "If you think process safety is expensive, wait until you have an accident."   In addition to regulations, societal and political pressures from the public demand ever-better safety and environmental performance.

Every company needs to find ways to improve its operating efficiency and performance, reduce overall operating cost, and at the same time find ways to maintain and improve its competitive market position. Improving market position and customer satisfaction is inherent in an organization's quality management program.   Although the management systems for process safety, occupational safety and health, environmental, quality and security may have developed separately, they have similar program-related expectations, such as being implemented with:

- Specific program-related record-keeping requirements, and
- Metrics used to demonstrate performance improvements of the program.

{*Note:* The management systems for process safety (S), occupational safety and health (H), environmental (E), quality (Q) and security (S) are sequenced for reference as "SHEQ&S" in this guideline.}

When the different SHEQ&S management systems are not well coordinated, the sometimes conflicting goals and demands on an operating facility may prompt program changes that inadvertently contribute to an increased process safety-related operating risk.   Unfortunately evidence of such conflicts exists today since industry still experiences many preventable incidents due to inadequate hazardous materials management systems and programs.   Examples include catastrophic equipment failures which resulted from inadequately designed, monitored and/or maintained equipment reliability programs. (Bloch 2012, US CSB 2003, and US CSB 2011b).

Other benefits for successful integration include reduced operating costs and more effective use of staff managing the programs, reducing duplication of effort across an organization. The history of successful business cost re-ductions is reflected in the improved results for organizations that im-plemented quality management programs.   Some of the benefits for integrat-ing programs using metrics which affect process safety performance and a quality management system approach are summarized in Table 1-1.   This guideline is written to address the need for integration between the process safety, occupational safety and health, environmental, quality and security management programs. Each of these programs has similar risk reduction goals that, once combined, will help a company become more efficient and effective when managing its overall operational risk.

Table 1-1. The "Business Case" for Process Safety

Business Value[1,2] - *Reduced incident costs*	
Ethical	Corporate responsibility
Employee	Fatalities, injuries, emergency response
Environment	Cleanup, material disposal, environmental remediation
Equipment	Repairs or replacement of failed component or damaged equipment as a result of subsequent fire or explosion
Financial	Flexibility, sustained value, business opportunity, business interruption, feedstock/product losses, loss of profits, obtaining or operating temporary facilities, obtaining replacement products to meet customer demand [e.g., from a sister facility at another location]
Business value[3] - *Integrating management systems across groups*	
Ethical	Distributed across the value chain and government entities and stakeholders
Community relations	Improved communications through Community Advisory Panels
Liability protection	Reduced insurance premiums, reduced terrorist liability [the Security Code meets Department of Homeland Security (DHS) requirements through the SAFETY Act as a Qualified Anti-terrorism Technology]
Organizational efficiency	Improve efficiency by taking advantage of and by combining existing management systems, encourages teamwork by bringing together diverse staff from multiple management teams (Groups: environmental, health, and safety; operations, maintenance, community relations; shipping; security; regulatory compliance; and purchasing)
Competitive advantage	Continuous improvement activity aligning environmental, health, safety, security, product stewardship and value chain performance
Business considerations[4] - *Fundamental principles*	
Humanist	Protecting the safety and health of employees and surrounding communities is the humanitarian thing to do - a company's moral obligation - regardless of legal obligation.
Employee / Labor relations	Employee involvement is a major tool in achieving quality safety and health. Consider areas in which employees can have a positive impact on safety performance.
Public Perception	Public perceptions about a company's attitude towards its employees can affect the market for its products.
Regulatory / Legal	Regulatory agencies aggressively enforce regulations; they can impose fines and cause operational interruptions. Companies and individuals may be held criminally liable for violations. The cost of litigating citations and proposed penalties against the company should also be considered. If found in violation, the company can lose some flexibility in how it allocates its resources. For uncontested violations, abatement must occur within the mutually agreed upon time period.
Financial	Consider the short- and long-term costs of adopting effective safety and health standards versus the increased cost of workers' compensation claims, lost time and other direct and indirect costs associated with a less effective program.

[1] CCPS, The Business Case for Process Safety, Second Edition, AIChE, 2006.
[2] CCPS, from the definition of "Direct Cost" in Process Safety Leading and Lagging Metrics, Revised: January 2011
[3] American Chemical Council (ACC), Business Value of Responsible Care©, http://responsiblecare.americanchemistry.com/Business-Value (accessed 18-September-2013)
[4] National Safety Council (NSC), 14 Elements of a Successful Safety and Health Program, (1994).

1.2   THE PURPOSE OF THIS GUIDELINE

One major goal of this guideline is to help an organization reduce its overall operational risk by integrating its monitoring-related work across groups, focusing on common high-risk metrics which affect process safety performance. T he purpose of this guideline is to present a process through which an organization could develop or improve thc ties between its existing process safety, occupational safety and health, environmental, quality and security management programs. Many metrics are common to more than one group, such that a well-designed and implemented integrated management system will reduce the work load on the process safety, safety and health, environmental, quality and security groups, and help improve manufacturing efficiency and customer satisfaction, as well.

The process described in this guideline uses parts of quality management approaches, such as Total Quality Management (TQM) or the ISO 9000/14000 series, providing an integrated management system that can be tailored to be consistent with a company's culture and management style (Albrecht 1990, ACC 2013b, Caropreso 1990, Juran 1964, Kane 1968, Scherkenbach 1986, Scholtes 1988).

1.3   THE SCOPE OF THIS GUIDELINE

The scope of this guideline focuses on the process for identifying common metrics between the process safety, occupational safety and health, environmental, quality and security management programs. S ince some of the metrics which affect process safety performance are common across groups and recent reviews on the types of process safety metrics have been published, this guideline has been written to capture the latest approach to help reduce an organization's overall operating risks. Although a quality management system may form the basic foundation for these risk management programs, it is beyond the scope of this guideline to detail the different types of quality management programs.

1.4   THE APPROACH USED IN THIS GUIDELINE

The existing business and SHEQ&S management systems that are integrated into the SHEQ&S program are shown schematically in Figure 1-1. For the purposes of this guideline, the "SHEQ&S program" is defined as the set of SHEQ&S management systems which monitor meaningful metrics to indicate process safety conditions. Metrics common to these groups are shown schematically in Figure 1-2, where the different SHEQ&S management systems have overlapping areas. Some metrics are common to different SHEQ&S groups, as is represented by the intersections in Figure 1-2. P lease note that the Safety systems include the two distinct process safety and personnel safety efforts essential for safe and reliable operations. The personnel safety efforts, in particular, are a part of the existing occupational safety and health programs.

Unfortunately, some metrics used for monitoring and tracking occupational safety and health programs have proven to be inadequate as the only measure for the real condition of the organization's process safety programs (see additional discussion in Section 1.8). Hence, the goal of this guideline is to help an organization identify the common metrics which affect process safety performance across the different SHEQ&S groups, as is represented by the "center" area of the intersection between management systems in Figure 1-2. When appropriate indicators are selected, tracked and monitored, an organization can reduce its overall operating risk across the different groups.

This guideline recognizes that companies may combine their risk reduction efforts into several different groups, with different combinations of the Safety (both process and occupational), Health, Environmental, Quality and Security groups (e.g., SH&E, HS&E, H&S, etc.). However, no matter what a company's organizational chart looks like, this guideline assumes that each group monitors group-specific metrics to ensure that its group's particular risks are reduced.

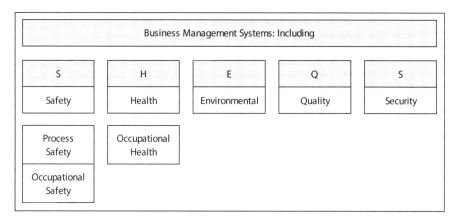

Figure 1-1. The Management Systems in the SHEQ&S Program

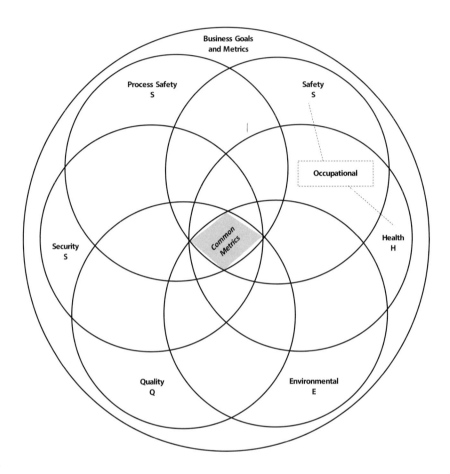

Figure 1-2. Metrics Common to the SHEQ&S Management Systems

The framework for organizing the material presented in each chapter combines the SHEQ&S program "Life cycle" phases and the "Plan, Do, Check, Act (PDCA)" approach as is shown in Figure 1-3. Each phase is briefly described below for Chapters 2 through 7:

Phase 1) The "Plan" intent for the SHEQ&S program:

The SHEQ&S program design begins at the initial "plan" phase (the program's creation or birth); with the understanding that reviews and gap analyses may change the program's design during its life as the programs mature and grow beyond their infancy.

The "Plan" phase chapters are:

Chapter 2. Secure Leadership Support across Groups

Chapter 3. Evaluate Hazards and Risks across Groups

Chapter 4. Identify Common Metrics across Groups

Phase 2) The "Do" intent for the SHEQ&S program:

The "do" phase is the day-to-day day application of each of the SHEQ&S systems. S uccess hinges on these systems being in place and adhered to by everyone, from those working in the field to those in senior management making decisions that affect the resources required to effectively implement the management systems. Safe, highly reliable organizations understand and apply the principles of conduct of operations and operational discipline.

The "Do" phase chapter:

Chapter 5. Implement the SHEQ&S Program

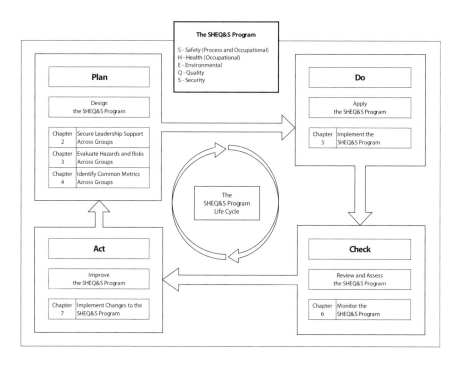

Figure 1-3. The Phases in the Plan, Do, Check, Act (PDCA) Approach

Phase 3) The "Check" intent for the SHEQ&S program

The "check" phase includes monitoring the SHEQ&S program metrics and auditing for trends. Every program needs to be reviewed on a regular frequency to ensure that organizational complacency does not occur.

The "Check" phase chapter:

Chapter 6. Monitor the SHEQ&S Program Performance

Phase 4) The "Act" intent for the SHEQ&S program

The "act" phase addresses the main driver of change for the SHEQ&S program: trending and gap analyses. New people, staffing reorganizations and findings from investigations or gap analyses may affect the selection of the SHEQ&S program metrics.

The "Act" phase chapter:

Chapter 7. Implement Changes to the SHEQ&S Program

For effective management of process safety risks within each SHEQ&S system, a company's culture and management style require strong operational discipline by everyone in the organization, whether they are contributing at the planning, doing, checking or acting phase, to help ensure and sustain safe and reliable operations.

1.5   HOW ESTABLISHED MODELS CAN BE USED IN INTEGRATED SYSTEMS

Different industries may manage Process Safety under various titles including Safety Management System (SMS), Operational Excellence (OE), Integrity Management Systems (IMS), Process Safety Management (PSM), Health, Safety & Environment Management System (HSEMS) or Security Management Systems (SeMS). Although there are different approaches and models that are tailored to meet a company's culture and management style, this guideline uses a structure that combines the CCPS's Risk Based Process Safety approach and international models (including the ISO 9000, ISO 14000 and the Certification Europe OSHAS 18000 series of standards) for illustrative purposes, recognizing that other management systems have similar structures. Additional management system frameworks are noted in the references at the end of this chapter. Whether a company is working with SMS, OE, IMS, PSM, HSEMS and/or SeMS systems, this guideline provides a methodology to help identify and select common metrics used to monitor and help improve process safety performance.

Some jurisdictions may require a "Safety Case" from which regulators expect the company operating a process with hazardous materials and energies to make the case for safety – the company has taken all measures necessary to prevent major incidents and have reduced their risk as low as reasonably practicable (ALARP). The Safety Case identifies the hazards and risks, describes how the risks are controlled, and describes the safety management system in place to ensure the controls are effectively and consistently applied. The basic principle is that those who create the risk must manage it. Because the company has the greatest in-depth knowledge of the hazards at its facility, the company must assess its processes, procedures and systems to identify its hazards, evaluate its risks and implement appropriate controls. This includes a demonstration that the company is employing recognized and generally accepted good engineering practices (RAGAGEP) in its engineering design, including human factors considerations, by using robust management systems. Although the Safety Case is not a management system, it demonstrates that a company complies with the regulation by having a safety management system included in its integrated SHEQ&S program.

1.6 EXCLUSIONS TO THE SCOPE

The scope of this guideline does not include advice on the development or implementation of specific business, process safety, occupational safety and health, environmental, quality and security systems and their respective programs. The guideline focuses on combining existing systems into an "integrated" SHEQ&S program based on common metrics which affect process safety performance. It is intended to provide a format, or a framework, for easy adaptation anywhere in the world. The references provided in this book provide multiple resources for detailing the design and implementation of the specific systems and their programs.

1.7 KEY AUDIENCE FOR THIS GUIDELINE

This guideline is intended primarily for people who help implement and monitor their group-specific risk reduction management systems, whether they are at the corporate, the facility or the process unit level of an organization. This includes the leaders and Subject Matter Experts (SMEs) within these groups: process safety, occupational safety and health, environmental, quality and security. This guideline will be a useful training tool and reference for corporate and/or site managers and leaders across all of the groups, helping them better understand the complexities inherent in reducing their overall operating risk (see discussion on developing leadership capabilities in Section 2.6). In addition, this guideline will help process safety auditors establish process safety-specific metrics that can be evaluated, both for program compliance and for system implementation at a facility.

This guideline applies to the people at small, medium and large facilities handling hazardous materials and energies, especially those required to have a formal regulatory or corporate-driven process safety management (PSM) program. The design of this guideline will benefit smaller facilities with limited resources, as well as larger facilities which struggle with inefficiencies across business units within the facility. Large corporations will benefit from integrated metrics when managing global corporate process safety risks, as well.

1.8   SOME RECENT ADVANCES IN PROCESS SAFETY METRICS

It is hoped that this guideline captures the essence of some recent advances in process safety metrics. Note that there are process unit-specific, facility-specific and company-specific metrics which apply to each group at each level. These metrics may not apply to the other groups or levels in the organization. In addition, it is beyond the scope of this guideline to describe in detail the different types of metrics which have been identified, such as "leading" and "lagging" indicators. Please refer to Appendix B for a brief overview and specific references for more details on recent advances in identifying and selecting process safety metrics.

2 SECURE LEADERSHIP SUPPORT ACROSS GROUPS

This chapter explores the need for securing leadership support at all levels across the SHEQ&S groups in the organization, addressing both the benefits and concerns which may be raised when proposing changes to the existing management systems. Not only is it important to have a vision of what the SHEQ&S program will look like and what the program's goals will be, it is important to emphasize the benefits and obtain support at the corporate leadership level first. With upper management support, a vision and the goals, the stakeholders at all levels in the organization will better understand their roles and how their group's resources will be shared and how they will benefit. Every group needs to understand their role in making the SHEQ&S program effective and how they will help improve the company's process safety performance.

This chapter introduces a case for the SHEQ&S program, addressing some of the group interactions and responses that can occur when external pressures force a crisis on an organization. The discussion on the process safety incident shows how an effective SHEQ&S program used to monitor metrics across each SHEQ&S management system could help an organization proactively monitor, respond to, and improve its process safety performance and manage its overall risk. Since part of the challenge when implementing a new program hinges on the capabilities of people to perform their new or enhanced roles, this chapter concludes with some references to help identify and address potential leadership competency gaps, improving the capabilities of process safety leaders.

2.1   THE NEED FOR SECURING SUPPORT

Securing support across the organization begins with visible support at the corporate leadership level. These leaders allocate the resources within their groups – the people, equipment and money – supporting operations at all levels in the organization. Without leadership and management support, no matter what level within the organization, the integration effort will be starved of resources and will most likely collapse.

The company's reporting or staffing structure is essentially the same no matter what its size: there are people who report to others in the organization all the way to the top chief executive officer (CEO), president or company board. This guideline will use the terms "corporate," "facility" and "process unit" to represent the three general levels in an organization. Although these three distinct levels are defined for this guideline, it is recognized that each organization may use different terms for these levels, such as some of the staffing terminology for the corporate level, the facility level, and the process unit level as listed in Table 2-1. Process safety leadership is expected by everyone in an organization, from those managing to those being managed.

Visible management support across all phases of a project is crucial for the success of any project. Since a typical project has several phases, the first phase, management support phase, is the most important. The other phases for the SHEQ&S program integration team's "project" include current system evaluation, conceptual design, detailed design, piloting, installation and testing, and then operation and maintenance, as is shown Figure 2-1. A project will not be effective or successful if it is not supported by those who allocate the resources at each phase of the project.

Table 2-1. Organizational level terminology

Organizational Level			Terms that may be used in an organizational chart
Corporate Level	Groups noted in this guideline		Process and Occupational Safety (S), Occupational Health (H), Environmental (E), Quality (Q) and Security (S)
	Other terms for this level: Enterprise Organization	Staff terminology	Includes president, vice president, executive, chief operating officer (COO), global director, global manager; includes global Process Safety Management (PSM) directors
		Regions	Includes Europe, North America, South America, Asia Pacific, Africa, Middle East
		Competency Centers	Includes process safety management (PSM), environmental, health and safety (EHS), engineering, maintenance, procurement, information services, supply chain, operations, operational excellence, research and development (R&D), sustainability
		Departments or Divisions	Includes financial, legal, taxes, insurance (loss prevention; property and casualty), strategic planning, communications, government relations, auditing, human resources, investor relations Divisions also noted with product-related groupings (e.g., chemicals, refining, upstream, downstream, etc.)
Business Level	Other terms for this level: Business Unit Business Stream Segments		A "business" is typically based on similar technologies or markets, such as refining, chemicals, specialty chemicals, advanced materials, biological, plant sciences, explosives, etc. Business Units may have facilities at different locations across the world
Facility Level	Groups noted in this guideline		Process and Occupational Safety (S), Occupational Health (H), Environmental (E), Quality (Q) and Security (S)
	Other terms for this level: Plant Site	Staff terminology	Includes facility manager, senior managers, assistant managers, deputy managers, engineers, officers; includes facility (site) PSM element owners
		Department terminology	Includes production, operations, maintenance, engineering, projects, quality control and assurance, information technology (IT), raw materials storage and/or warehouse, purchasing, customer service, human resources, administration, accounting, finance
Process Unit Level	Other terms for this level: Assets	Staff terminology	Includes operators, mechanics, electricians, technicians, process support engineers, laboratory technician, attendants, workers, line supervisor; includes local PSM element owners
		Hazardous process terminology	Processes that handle hazardous materials and energies with the potential for harm to people, the environment and property if the equipment designed to control them fails; consequences: fatalities, injuries, environmental and property damage resulting from toxic releases, fires, explosions, and/or runaway reactions

Everyone should understand and support the SHEQ&S program as a part of the organization's overall management system, with the SHEQ&S program eventually becoming the normal work process in the organization. Answering the following questions will help explain the rationale when the program is being introduced:

- Who benefits from the SHEQ&S program?

- What are the benefits of the SHEQ&S program?

- What will the final SHEQ&S program look like?

- How does the SHEQ&S program differ from the current systems?

- How will the change be achieved?

Potential answers to these questions for the different stakeholders are shown in Appendix C. Effective communication must occur between those owning and doing the work and management supporting the work as the SHEQ&S program moves from its conceptual stage through its piloting stage and then to its implementation stage.

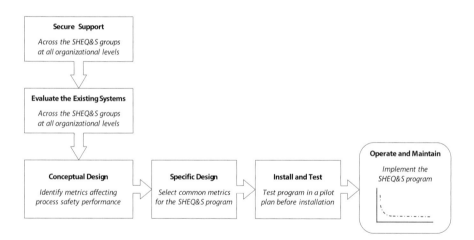

Figure 2-1. Typical phases in a project: Focusing on the SHEQ&S program

2.2   SECURING SUPPORT TO OPTIMIZE RESOURCE ALLOCATION

An effective SHEQ&S program helps manage the company's overall operational risk, which can be represented by expanding the general risk equation into an overall company risk equation, as is shown in Figure 2-2 and represented with the risk matrix in Figure 2-3. The overall operational risk is a function of the event's frequency, its consequences, the company's resource allocations, and the company's operational discipline. The goal is to manage the company's operational risk to its lowest tolerable level by implementing systems to help reduce adverse event frequencies, to help reduce their consequences, and to help increase operational discipline [CCPS 2011].

However, when adding "resource allocation" to this equation, the overall company's efforts become complicated. From a process safety risk management perspective, the danger in being too focused on the more routine, high frequency, low consequence events is that the low frequency, high consequence events may be overlooked, and therefore be unmitigated [Murphy 2011, Murphy 2012, Murphy 2014].

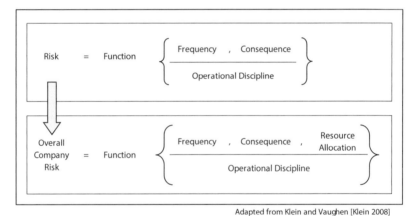

Adapted from Klein and Vaughen [Klein 2008]

Figure 2-2. A general equation for a company's overall operational risk

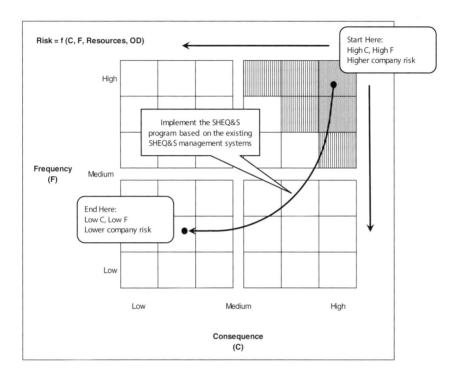

Risk = f (C, F, Resources, OD)

Start Here:
High C, High F
Higher company risk

High

Implement the SHEQ&S
program based on the existing
SHEQ&S management systems

Frequency
(F)   Medium

End Here:
Low C, Low F
Lower company risk

Low

Low   Medium   High

Consequence
(C)

Figure 2-3. A general risk matrix for a company's overall operational risk

The benefits of an effective SHEQ&S program include ensuring that these low frequency, high consequence process safety events do not occur due to misunderstandings, miscommunications, or organizational loss of memory over time. Ineffective focus within an organization may place the organization at greater overall operational risk by allocating too many resources to address one group's relative risk at the expense of adequately resourcing and addressing the risks affecting another group. The increase in overall risk is shown schematically in the risk profile/resource allocation chart in Figure 2-4. T he optimum, appropriate risk mitigation occurs in between the liberal and conservative limits, with too few resources on the left and too many, potentially ineffective resources on the right. A company has a limited number of resources which must be allocated effectively to obtain its appropriate risk mitigation level; it cannot afford to divert its resources on efforts that do not keep its overall risk management efforts towards the center of the curve. Recognizing that there are other risks that extend beyond those addressed within the different SHEQ&S groups, including its business and societal risks, it is important to note that each organization must address their risks effectively to remain in business.

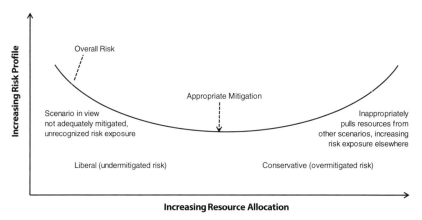

Adapted with permission from the Phillips 66 Company

Figure 2-4. Assessing overall risk to ensure appropriate resource allocation

An example of greater operational risk occurred when a co mpany's environmental volatile organic (VOC) emissions reduction efforts did not adequately address the process safety hazards and risks. The company responded to new environmental permitting issues on reducing its volatile organic compound (VOC) emissions by adding an ethylene oxide incinerator. U nfortunately, the incinerator ignited the ethylene oxide discharge stream, causing a flash back and subsequent explosion. Four employees were injured and the operation was shut down for almost six months [US CSB 2006]. A n effective SHEQ&S program could be designed to leverage resources across groups to help a company reach an "appropriate operational mitigation" level of risk across the different groups.

2.3   DEVELOPING A PRELIMINARY PLAN

Before the integrated plan can be developed, its "vision" must be developed first. This vision is our destination. T he plan is the journey. What will the fully integrated system look like? A preliminary image portraying this vision and helping to answer this question, is shown in Figure 2-5, with the existing state of the separate SHEQ&S m anagement systems on the left (the "current state") contrasted with the SHEQ&S program on the right (the "future state"). S ince different companies will have their own management structure, the general diagram in Figure 2-5 can be tailored with the goal of integrating each of the separate SHEQ&S systems into one effective SHEQ&S management system that interacts with the overall corporate management system (e.g., for the current state, a company may have "EHS" as one of its blocks).

In addition, there may be other existing management systems that a company has which would help factor into the "future state" of the SHEQ&S program, such as management systems that address loss prevention, property and casualty insurance, and workers compensation costs which inevitably result after a significant process safety incident. S ustainability and product stewardship programs, addressing a company's "green" efforts or its "cradle to grave" responsibilities, could be considered in the integration effort as well. Considering these other management systems, this guideline specifically focuses on and discusses the SHEQ&S groups, recognizing that each company should define which management system it plans to integrate into the final system.

In many companies, the process safety and environmental groups and the safety and health groups may have developed independently with their own set of management processes, programs, elements and resources. People who have been through audits, whether on process safety, safety engineering, occupational safety and health, environmental, technical engineering, loss prevention, insurance, quality or security management systems know that similar if not identical questions are asked across the facility's programs and elements (e.g., incident investigations, emergency responses, documentation and record keeping, etc.). This duplication of effort, from both the corporate auditing perspective and the facility's resourcing perspective, is not value-adding and can be eliminated with an effective SHEQ&S program.

By integrating similar regulatory requirements common to the process safety, occupational safety and health, environmental, quality and security groups, the consolidated, streamlined system integrates the staffing resources required to manage them. T he integration will provide a more flexible system capable of incorporating changes more quickly. Hence, the resulting system provides a more effective metrics "dashboard" that better conveys changing business and regulatory requirements.

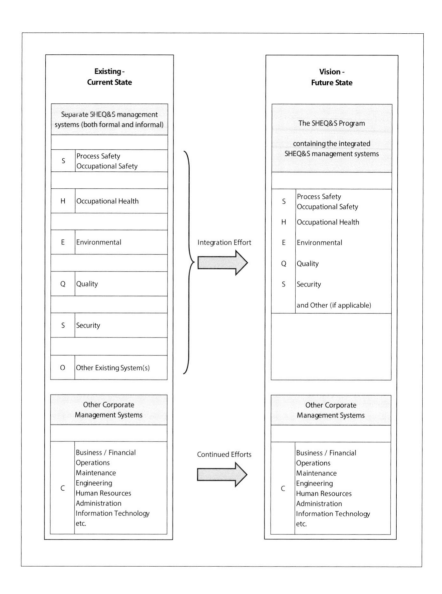

Figure 2-5. A preliminary vision for answering the question "What will the final system look like?"

Depending on the organization's SHEQ&S management system maturity level, additional supporting material may need to be compiled and used for justifying this effort. A process for creating the preliminary plan could include the following steps:

1) *Inventory* current SHEQ&S management systems and elements.
   - Chapter 3 of this guideline describes an approach for developing a detailed inventory
2) *Develop* initial listing of systems to be integrated across the groups.
   - This is addressed in Chapter 3, as well
3) *Identify* metrics which affect process safety performance for the integrated system.
   - Chapter 4 of this guideline describes a risk ranking approach to help identify these common metrics
4) *Demonstrate* usefulness for the SHEQ&S program.
   - This is illustrated with the case study in Appendix D. When piloting the SHEQ&S system, see the discussion in Chapter 5
5) *Evaluate* whether decisions from other corporate management systems may have an overriding effect on the SHEQ&S program (e.g., limit availability of resources which managers within the integrated system do not control).
   - This evaluation effort is beyond the scope of this guideline, but an organization must identify its weakest barrier (its weakest layer of protection) within its overall organizational management system. It is not practical to measure all systems and the priority for an organization's survival must be focusing on its weakest management system.
6) *Integrate* the organization's SHEQ&S management systems in context of this guideline. This helps an organization become more effective in managing its risks and improving its process safety performance.

Examples of successfully implemented SHEQ&S programs that address the resource allocation issues noted in Section 2.2 have been provided in Chapter 8.

2.3.1   Addressing some of the benefits and concerns

The next three sections discuss costs and benefits which need to be addressed before integrating the different groups into a SHEQ&S program. Section 2.3.2 explores some of the SHEQ&S program benefits. Section 2.3.3 addresses some of the management concerns (i.e., its "costs") and Section 2.3.4 addresses the organization's benefits from an overall management perspective. Each of these sections concludes with a table summarizing some of the SHEQ&S program's related benefits and concerns (i.e., Table 2-2, Table 2-3, and Table 2-4).

2.3.2   Addressing some of the SHEQ&S program benefits

In addition to improvements in efficiency and effectiveness, as discussed earlier, other possible benefits include lower operating costs, enhanced problem solving, more effective standards and procedures, active continuous improvements, useful measurements with sound statistical analyses, and satisfied customers within and external to the company. S ince these approaches are designed considering a typical quality management system, it is natural to borrow from the quality group's systems and apply them to the other groups, as needed. The particular benefits for a SHEQ&S program are summarized in Table 2-2, which may be useful when justifying and securing support for the resources to charter an integrated SHEQ&S team.

2.3.3   Addressing some of the management-level concerns

Developing, designing and achieving a SHEQ&S program will not be without some initial cost in resources, which will depend on the maturity of the existing systems from which to work. Using a risk based approach adds value when it makes sense, rather than applying the same approach across the high risks as well as the low risks [CCPS 2007a]. Securing support across the organization, whether at the corporate, facility or process unit level, will hinge on how well management addresses safety concerns in context with the other costs and concerns. Common concerns include high implementation costs, apprehension that the final system does not achieve the cost reduction goal, trepidation that important issues will be missed, and most worrisome to some, a "loss of control" since part of their group's specific performance improvements cannot be controlled directly by their group any more. The particular concerns focusing on a SHEQ&S program, including its potential implementation costs, are summarized in Table 2-3, which can be used when securing support for the SHEQ&S program effort across the organization.

Table 2-2. Potential SHEQ&S program benefits

System benefit	Quality management approaches	Benefits when managing process safety
Lower costs	Enables management to reduce both the costs associated with services and the costs of consequences.	Helps reduces overall process safety-related services and reduces the costs associated with process safety incidents.
Improved problem solving	Requires everyone at all levels in the organization to be a part of the development of a solution.	Can help identify and verify that the process safety-related risks are not missed by those proposing a change, especially on process equipment critical to process safety. By following the management of change procedure and completing, if needed, the equipment installation pre-startup safety reviews, all affected groups will understand the proposed change.
Work process consistency	Requires written standards and procedures with scheduled, documented reviews and authorizations.	Helps in process safety system audits, which expect written process safety-related standards, written process-related "critical" operating procedures and written process safety-related "critical" equipment maintenance procedures that are controlled, reviewed and updated at specific frequencies. Some of these procedures may be identified and fall under specific regulatory review frequencies, as well. Helps with SHEQ&S-related training efforts by integrating the common hazards-related knowledge across group-specific training packages (e.g., design and use of a training matrix).
Continuous improvements	Requires that continuous improvement be built into the processes. Better efficiencies and higher quality is always possible.	Helps employees look for ways to improve both process-related efficiencies and help improve process safety performance. Depending on what gaps are identified as the baseline, significant improvements usually occur within a short time, followed by continuous steady improvement.
Clearly identified measures	Continuous improvements depend on measurements of key process parameters for tracking progress (or lack thereof), helping identify and proactively correct deficiencies before they become a crisis that results in a large consequence.	By using the different types of process safety performance indicators, such as "lagging" or "end of pipe" indicators, "leading" or "dash board" indicators, and process safety system efficiency indicators, process safety performance improvements are effectively tracked.
Sound statistical data analysis	Statistical concepts are used to analyze the measurements. With proper selection of the type of measurement (i.e., discreet, continuous) and the type of analysis (trending, benchmarking, etc.), a performance baseline can be measured and, with time, show tangible improvement.	Statistical concepts are used to analyze the performance indicators. With proper selection of the type of measurement (i.e., discreet, continuous) and the type of analysis (trending, benchmarking, etc.), a performance baseline can be measured and, with time, show tangible improvement. A smaller, manageable number of indicators are selected, focused upon and responded to. Hence, people are not overwhelmed by the larger number of indicators which exist when every SHEQ&S group has its own set of indicators.
Satisfied and engaged "customers"	Requires a deliberate effort to identify all customers and satisfy their expectations.	Using a process safety lens, the ultimate "customer" expects the process safety-related efforts to protect them from the hazards. These customers are internal, the employees working at the facility, and external, the people living in surrounding communities.

Table 2-3. Potential management-level concerns

Management concern	System-related concern	Potential resolution for the concern
High implementation costs	Potential costs include allocating resources for the team needed to design the integrated SHEQ&S system, possibly hurting performance during the design, piloting and implementation efforts as the organization becomes more familiar with the integrated system.	Determine how other corporate-level initiatives have fared in the past, glean from their implementation failures and successes, and incorporate the successes into the proposed integrated SHEQ&S pilot. Since there is no prescriptive success formula, short terms costs will occur. However these initial costs will be far overshadowed by the long term benefits to the company.
Does not achieve cost reduction goal	Fear that the proposed system change, the integrated SHEQ&S system, will not succeed in reducing costs.	By design, the integrated SHEQ&S management system is risk based. It will have fewer resources tracking and responding to the prioritized metrics affecting process safety performance. These metrics will be effectively measured and tracked only once, removing this monitoring and tracking responsibility from the other groups, helping reduce costs.
Misses important issues	This concern depends on the maturity of the existing systems, especially if the system has been around for a long time. Older more established systems should have already identified and corrected major gaps through audits and findings. Newer systems may not have had the time to do so.	The risk based process safety approach identifies the important process safety performance metrics and the existing formal and informal systems used to monitor and track these metrics across the groups. Gaps are addressed in the piloting and implementation efforts for each SHEQ&S group.
Loss of control (affected by decisions from another group)	Divisions in the reporting structure in an organization complicate the integration effort when one group's specific risk reduction goal conflicts with one or more risk reduction goals from another group.	Determine the day-to-day demands for the processes handling the hazardous materials and energies. By working up the reporting structure, the staffing between groups, no matter what level, can be identified. The integration team can identify and address resource- and risk-related gaps and conflicts by using the management system assessment tools.

2.3.4 Sharing some of the management-level and company's benefits

Most managers don't need much persuasion for supporting new programs that show personal benefits along with the company benefits. Hence, for the integrated SHEQ&S vision to gain support, the following benefits should be shared from the "personal" management perspective, as well: 1) there are fewer processes to manage; 2) the time on process safety, occupational safety and health and environmental issues is more effectively managed within the SHE groups; 3) there is more effective change management; 4) there is better performance measurement from which to base decisions; and 5) continuous improvement of the work process becomes a way of life (additional discussion for continuous improvement when effectively managing process safety is noted in the Risk Based Process Safety (RBPS) approach [CCPS 2007a]). The manager and organizational benefits for a SHEQ&S program are summarized in Table 2-4.

Table 2-4. Potential management-level and company benefits

Benefit	Use of the SHEQ&S program
Fewer processes to manage	The SHEQ&S program integrating the SHEQ&S management systems will require less management attention across the organization.
Time on safety, health and environmental issues	The SHEQ&S program will help managers more effectively manage the time spent on process safety, occupational safety and health and environmental issues. For managers in other groups, "process safety" is a small part of their total responsibilities. Hence, the integrated SHEQ&S management system will provide more time for the managers in other groups to focus on the needs affecting their group.
More effective management of change	The SHEQ&S program is consciously built to be adaptable to inevitable changes. Changes will be more effectively managed to help identify and reduce associated process safety-related risks when new safety, health and environmental regulations occur, when process technology is updated, or when equipment-related continuous improvement efforts are proposed.
Better performance measurement	The SHEQ&S program will proactively identify gaps focused on improving process safety performance, therefore preventing potential harm to people, the environment and to property. Risk-based indicators will be identified across groups. They will be effectively monitored, tracked, monitored, and evaluated, with the gaps addressed when needed.
Better external (tolling) management	The SHEQ&S program incorporates a product stewardship ("cradle to grave") philosophy which can be extended to external, tolling organizations providing a service to the company. Specialized, contracted warehousing operations managing a company's materials must understand the hazardous materials and how the company manages the associated risks.
Continuous improvement	The SHEQ&S program has continuous improvement designed into its management process. No successful organization has survived in the long run if it does not address the changing world and its external influences on the overall performance of the organization.

2.4   THE IMPORTANCE OF A SAFETY CULTURE

In addition to leadership support, a strong safety culture is essential for the success of a SHEQ&S program. A strong and healthy process safety culture helps prevent injuries, saves lives, and improves productivity. The UK Health and Safety Executive defines safety culture as "…the product of the individual and group values, attitudes, competencies and patterns of behavior that determine the commitment to, and the style and proficiency of, an organization's health and safety programs" [HSE 2002]. A more succinct definition has been suggested: "Safety culture is how the organization behaves when no one is watching" [CCPS 2015].

A good safety culture allows people to question the current engineering and administrative controls, recognizes and resists complacency, commits to excellence, and fosters both personal accountability and corporate self-regulation.

People in a highly reliable organizations understand how to effectively respond to uncertainty at the time the incident is occurring, recognizing the hazardous material or energy threats and acting safely with the values ingrained by the way things are done. Since it is beyond the scope of this guideline to address how people best achieve their goals through purposeful and safe behavior, the reader should consult other references for understanding the need for a strong and healthy process safety culture across all levels of a highly reliable organization [ACS 2013, Bond 2007, CCPS 2015, Ciavarelli 2007, Dekker 2007, Gunningham 2011, HRO 2013, Koch 2007].

In summary, a process safety culture is simply inherent in a safe and reliable organization. A strong and healthy safety culture reflects the actions, attitudes, and behaviors of everyone in the organization with process safety as a core value. Process safety is simply a part of the way work is done.

2.5   IDENTIFYING STAKEHOLDERS

The "stakeholders" for the SHEQ&S program are those who are affected by and benefit from the integrated system. Their needs and concerns must be identified and addressed for the system to effectively reduce the work load across the organization. From the quality management perspective, the stakeholders are the suppliers and the customers; for the SHEQ&S program, the supplier is essentially the group measuring a specific process safety metric; and the customers are those using the metric from which decisions are made.

In the case for improving process safety performance, the stakeholders who have different needs yet have an interest in and benefit from safe facility operations include:

- Employees – who expect to be provided with a safe workplace, including operators, maintenance workers, laboratory technicians and technical staff (includes PSM element owners at all levels in the organization).
- Contractors (internal; located at a facility) – who expect to be provided with a safe working environment and understand their role in achieving this.
- Contractors (external; tolling operations providing a service for the company under a contracted fee) – who are expected to understand the processing and material hazards when using equipment at their facilities.
- Managers – who want easy-to-use and effective management systems that cover all the process safety, occupational safety and health, and environmental issues.
- Owners – who don't want the value of the company to be harmed by poor performance.

- Neighbors – who need to be assured that they will not be harmed by the operations.
- Local politicians and community leaders – who may welcome the employment in the community but may concerned that the risks may be too high.
- Regulators – who expect compliance with all regulations and standards.

The process unit level, the facility level and corporate level represent the series of suppliers and customers within an organization, with the ultimate customer being the overall organization. The process unit metrics are used at the process unit level, with the compiled and aggregated metrics used at the facility level, and the compiled and aggregated facility metrics used at the corporate level. Considering this approach, the final "product" is improved process safety performance due to effectively implemented SHEQ&S program monitoring metrics.

The activities or management processes along the "customer chain" consist of information, raw materials, equipment, and/or products. I n particular, the hazardous materials and the equipment or assets required to contain them are located at the process unit level. T he facilities may contain one or more process units and are geographically bounded by fences, their "fence line," from the surrounding areas, whether populated or not. The company may contain one or more facilities across the world, each with process units that have to control hazardous materials and energies. A schematic of these three levels is shown in Figure 2-6, noting that the "business unit" with similar technologies and hazards may complicate this organizational structure by having process units distributed across different facilities.

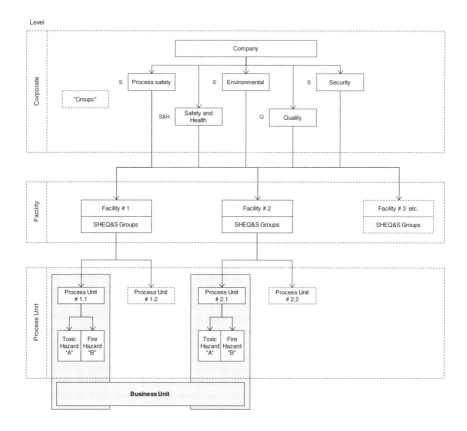

**Figure 2-6. The general organizational structure
for the terms used in this guideline**

The stakeholders for the SHEQ&S program include the people located at every one of the levels, including those dedicated to specific business units (these are "internal" stakeholders). The customers from the process unit level to the corporate and business levels are relying on effective management of the hazardous raw materials and products, on the integrity of the equipment designed, fabricated, installed, operated, maintained and changed to control the hazardous materials and energies, and on the effectiveness of the process safety systems and programs designed to manage the hazards. The owners, neighbors and local communities are essentially "external" stakeholders who rely on the hazards remaining on the inside of the fence.

2.6 SHARING RESOURCES ACROSS GROUPS

There are four general resources used at all levels in an organization:

- Human – staffing, with both knowledge and skills
- Physical – equipment, materials, parts, and land
- Financial – budgets for capital, services and other expenses
- Information – for decision-making at all levels of the company.

Upper management must drive sound cultures. Important cultural values include safety, environmental stewardship, ethics, and respect for others. People need to be clearly shown what is wanted. Everyone needs to know what road they are on and what their role in reaching the destination is (the "vision"). If people are provided the right tools to do their job, they will be able to do it well. If people have effective and efficient systems that help them get the job done, with adequate information to make decisions and adequate funding, a company can meet all of its objectives.

2.7 THE CASE FOR A SHEQ&S PROGRAM

A detailed case study has been developed for this guideline to help illustrate the case for designing and implementing an effective SHEQ&S program. This case study is based on an incident that resulted in a contractor fatality [US CSB 2011a], expands upon the CSB report, and explores a cause which may have contributed to the incident. In particular, the case study identifies management decisions across the organization over time that adversely affected the operating and maintenance stages of the equipment life cycle. To paraphrase Trevor Kletz: "You may disagree with what I think is the cause, but the incident happened and we need to do something different and make sure it doesn't happen again." The incident did occur, resulting in a contractor fatality.

The case study explores how cuts in an equipment preventive maintenance program contributed to the incident, noting that such cuts have been cited as a causal factor in significant incidents, resulting in fatalities, injuries, severe environmental harm and significant property damage. J ohn Bresland, then Chairman of U.S. Chemical Safety and Hazard Investigation Board, stated in 2009:

"Even during economic downturns, spending for needed process safety measures must be maintained ...companies should weigh each decision to make sure that the safety of plant {process unit} workers, contractors, and communities is protected. In the long run, companies that continue to invest in safety will reap benefits far into the future [CCPS 2009b]."

The equipment operating and integrity framework for this case study is based on the equipment life cycle which is described briefly in Appendix E. Decisions made by different operating groups affected many of the life cycle stages of the equipment, including its operation and maintenance stages. As is described in the CCPS RBPS guideline, asset integrity and reliability are essential elements for maintaining equipment and for effectively managing process safety risk [CCPS 2007a, Sepeda 2010].

The case study shows how an effectively designed and implemented SHEQ&S program helps identify process safety gaps proactively, allowing time for addressing and correcting issues before it is too late. The detailed case study uses the RBPS approach to help describe the increased process safety risk over time. In particular, the case study illustrates how the organization's overall operating risk increased when the equipment was not operated within its design specifications after not being maintained per its preventive maintenance schedule. The details for this case study are presented in Appendix D.

2.8   SURVEYING FOR COMPETENCY GAPS

Visible management support and personnel competency across all levels in the organization is crucial for the success of the integrated SHEQ&S program. If competency gaps are not identified and addressed early, the design, piloting and implementation efforts of the integrated SHEQ&S management system will be jeopardized. This section provides an overview of two surveys developed to help an organization evaluate for potential gaps in leadership competency and to evaluate for potential gaps in the existing management systems. This evaluation phase is the second phase shown in Figure 2-1 in the overall SHEQ&S integration "project," helping an organization evaluate and identify potential gaps *before* designing its integrated SHEQ&S management system.

The two evaluation surveys are framed using the Risk Based Process Safety (RBPS) management system guidance provided by the CCPS [CCPS 2007a, Sepeda 2010]. There are twenty pillars identified for a successful management system based on the following foundations:
1) Commit to process safety
2) Understand the hazards and risks
3) Manage risk
4) Learn from experience.

For both surveys, the RBPS framework is listed in the rows with the different SHEQ&S or process safety-related responses listed in the columns. The questions in the surveys help identify potential management system and potential personnel competency gaps.

These surveys are provided in Appendix F and in Appendix G, with a brief description of each survey noted below.

Appendix F: the "SHEQ&S Management System Mapping Survey"

The premise is to-
Successfully reduce the work demands on the different SHEQ&S groups by understanding and enhancing the existing management systems, not creating new work processes.

The questions posed in the SHEQ&S system mapping survey focus on the systems used to manage an organization's operational risk across the SHEQ&S groups. Since global organizations have facilities under different jurisdictions and regulations, its corporate standards and guidelines must be performance based, allowing each facility to develop their prescriptive, facility-specific standards and guidelines.

Appendix G: "The Process Safety Personnel Competency Survey"

The premise is to-
Successfully implement the integrated SHEQ&S management system with competent personnel across all levels in an organization.

The questions posed in the process safety competency surveys focus on the personnel applying the corporate and facility's process safety-specific management systems. Gaps in personnel accountability, if any, are identified quickly, helping ensure that everyone knows what their role is, from those responsible for providing the resources to execute the corporate or facility programs to those responsible for executing the design, construction, operation or maintenance of the equipment in the field.

These surveys are designed to evaluate the management systems and personnel competencies across each of the SHEQ&S groups at all levels in the organization. For successful design, implementation and sustainability of the integrated SHEQ&S system, gaps in the process safety-related roles and accountabilities identified with these surveys must be understood and addressed.

3 EVALUATE HAZARDS AND RISKS ACROSS GROUPS

This chapter addresses the need for establishing a common process safety-related risk reduction foundation, focusing on the process unit-specific hazardous materials and energies and their risks that have the potential for hazardous consequences: toxic releases, fires, explosions and runaway reactions. However, each SHEQ&S group has other specific group-related hazards with group-specific risk reduction efforts, as well, as was shown with metrics that do not overlap the other groups in Figure 1-2 (Chapter 1).

It is important to remember that this guideline focuses on process safety-related hazards; in particular, the metrics which affect process safety performance selected for the SHEQ&S program that can have, in some part, a measurable influence on and help improve the performance in the other groups, as well. Ultimately, it is the people on the front line, those in the process unit during its day-to-day operations, who have the most direct influence on the process safety metric, with guidance provided by those in the process safety group.

3.1   THE NEED FOR EVALUATING HAZARDS AND RISKS

For a company to remain competitive and be successful, it needs to understand the hazards and risks to the enterprise (its overall operations) and responsibly manage its operating risks, continually evolving its strategies as internal and external demands change. The corporate-level process safety risk strategy defines its tolerable process safety risk, expecting operations at each facility to implement management systems to control and manage the hazardous materials at each of its process units. A potentially high consequence event could destroy the enterprise if it is not properly addressed at the process unit level. A general company risk matrix was shown in Figure 2-3, with the goal for implementing engineering and administrative controls to reduce the risk to a tolerable level, then manage the risk of operations.

The decisions required by an enterprise to survive and remain competitive are often difficult to make, as the interactions between each SHEQ&S group is complex and includes interactions with other essential company-related groups, such as legal, business/finance, and human resources. If a company does not properly understand these complex interactions, implementing any one group's top risk reduction measure without considering the efforts of another group may increase the overall operational risk to an unacceptable level. Unfortunately, past history in many industries, including the chemical and refining industry, as well as the aviation and nuclear industries, continue to have significant incidents that have caused fatalities and significant environmental harm when the decision to reduce operational risk in one group did not adequately address the impact across the entire organization.

3.2   IDENTIFYING AND PRIORITIZING KEY PROCESSES AND RISKS

Management systems such as ISO 14000 r equire companies to identify key processes and associated risks. Most companies have developed key performance indicators that measure the health of each SHEQ&S group's management system, helping them prioritize resources to address gaps in each group's performance. After the Buncefield incident, the UK HSE began to require its chemical industry to link "warning sign" metrics relative to the location of and the risks at the facility [HSE 2008, HSE 2011a]. H ence, systems designed to manage a company's process safety risks, focusing on controlling the process unit's hazardous materials and energies, measure the performance of both their preventive and mitigative barriers. The difference between these warning signs (the leading indicators) and incidents (the lagging indicators) was described earlier in Section 1.8.

The facilities with process units handling hazardous materials and energies which could lead to toxic releases, fires and explosions are the key facilities for the SHEQ&S program. Each company must understand its potential process safety-related consequence, both within the site as well as to the surrounding communities, and determine its level of tolerable risk.

To help identify and prioritize the process safety risks, in particular those associated with the loss of containment of hazardous materials and energies, a Bow Tie diagram (Figure 3-1) can be used. The threats on the left affect each SHEQ&S group, with all of the threats to the other groups corresponding to threats to the process safety group. The preventive barriers are specifically designed to prevent a loss of containment of hazardous material or energy, whereas the mitigative barriers on the right are designed to reduce the consequences of such a loss of containment.

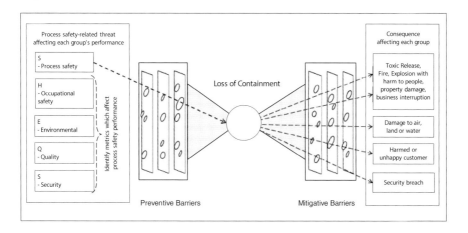

Figure 3-1. The Bow Tie diagram as a framework for helping identify metrics which affect process safety performance

3.3 SELECTING POTENTIAL METRICS

One of the goals of this guideline is to help an organization select from the many different metrics which affect process safety performance to ensure its lowest overall operational risk. Once the key processes and their associated hazards and risks have been identified, the key metrics can be prioritized and selected. Each group measures its group-specific metrics to some extent already; the integrated exercise described in more detail in Chapter 4 can be used to establish a common language between groups.

In general, the common language between the SHEQ&S groups (the common, overlapping metrics shown in Chapter 1, Figure 1-2) can be represented by the matrix in Table 3-1. Based on this table, there are common metrics which could be identified and used for the SHEQ&S program. There are common leading and lagging indicators which measure the health of the systems designed to control the hazardous materials and energies. These indicators measure the operational risks and help protect people, the environment, and the assets. The leading metrics could identify adverse effects to product quality, environmental permit deviations, or potential loss of containment incidents. The lagging metrics could identify the consequences of the loss of containment, measuring injuries or fatalities, harm to the environment, or asset damage due to toxic releases, fires or explosions.

Table 3-1. Potential intersections between groups for common metrics which affect process safety performance*

SHEQ&S Group		Metrics affecting process safety performance: Addressing the potential harm to life and health						
		Leading Indicators (Note 1)			Lagging Indicators (Note 2)			
		Operations	Maintenance	Change	Loss of Containment	Toxic Releases	Fires	Explosions
S	Process safety	✓	✓	✓	✓	✓	✓	✓
H	Occupational safety	✓	✓	✓	✓	✓	✓	✓
	Occupational health	✓	✓	✓	✓	✓	✓	✓
E	Environmental	✓	✓	✓	✓	✓		
Q	Quality	✓	✓	✓	✓			
S	Security (Note 3)	✓			✓	✓	✓	✓
Number of groups with metrics which may affect process safety performance		6	5	5	6	5	4	4

Details for the potential process safety performance metrics that can be chosen for each hazard/group are provided in Chapter 4.

Some examples of leading metrics include the following:

- Operations - measuring deviations from the standard operating limits, exceeding the process and equipment design conditions;

- Maintenance - measuring deviations from standard equipment tests and inspections;

- Engineering - measuring deviations from the original equipment design intents.

Some examples of lagging metrics include:

- Near misses – measuring incidents that, had conditions been slightly different, the consequences would have been more severe (i.e., had it happened at a different time, more people would have been hurt);

- Loss of containment – measuring activation of relief systems (exceeding safe equipment design) and measuring leaks (small and large).

3.4   FOCUSING ON PROCESS SAFETY PERFORMANCE

Once the metrics have been identified at all levels in the organization, it is essential that improvement goals are set, periodically reviewed, verified and updated as the program evolves, and that progress is shared across the organization. Although it is important that process safety performance success be monitored and shared, it is essential that weaknesses and deficiencies as well as the gaps in the process safety systems be identified and corrected.

A successful process that helps ensure that the performance gaps are monitored and that improvements are tracked includes these leadership aspects at the corporate, facility and process unit levels:

- Clear and visible commitment to improving process safety performance
- Clearly identified and tracked process safety performance metrics and gaps
- Clearly assigned responsibility for those managing process safety, and
- Clear accountability for those responsible for specific process safety-related efforts to show continuous improvement over time on their specific metrics.

Although each SHEQ&S group will be tracking process safety-specific metrics within the framework of their group's programs, there is guidance on a general process for using metrics to drive performance improvements (Chapter 7 and CCPS 2010). The goal is to effectively monitor and track process safety performance improvements through the SHEQ&S system using common *and* group-specific metrics which provide a measure of an organization's process safety performance.

3.5   RE-EVALUATING METRICS FOR CONTINUOUS IMPROVEMENT

However good the initial design of the SHEQ&S program is, opportunities for continuous improvement will always occur. I n addition, changes to the regulations, to the process units, to the staffing, and to the management processes will occur over time. As our knowledge of process safety, occupational safety and health, environmental, quality and security hazards continually improves, we add new techniques to our group toolboxes, as well. Continuous improvement does not happen accidentally; it requires a deliberate attempt to include it in the overall process. The framework developed must contain all the elements of continuous improvement, including a feedback loop to assure that the proposed changes do not introduce any new problems or issues with the other groups. Hence, building continuous improvement into the management process is vital if the integrated system is to withstand the test of time.

There are several elements to continuous improvement which should be included in the overall framework of the SHEQ&S program PDCA life cycle introduced in Chapter 1, Figure 1-3. The basic PDCA structure ensures that any deficiencies in the existing system will be identified and corrected across all groups. This includes steps in the management process that identify what activities need to be done (plan), apply the activities (do), measure and/or review the activities to make sure the system is working as expected (check) and implement changes to correct any problems (act). While not all these steps will be active all the time, the overall management process should ensure that all the continuous improvement activities take place on a regular, periodic basis.

Since continuous improvement requires a feedback loop to assure that the proposed changes do not introduce any new problems or issues, another way to view the effort is through the PDCA life cycle lens shown in Figure 3-2. The following phases are summarized below, beginning at the "You Are Here" arrow and returning to the same starting point (the continuous improvement cycle):

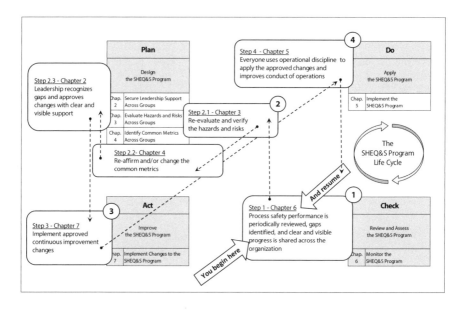

Figure 3-2. Applying continuous improvement efforts throughout the SHEQ&S program life cycle

Phase 1 – Check

Monitor the SHEQ&S Program Performance (Chapter 6)

Process safety performance is periodically reviewed, process safety system gaps are identified, and clear and visible progress of implemented improvements is shared across the organization.

Phase 2 – Plan

Phase 2.1 - Evaluate Hazards and Risks across Groups (Chapter 3)

The process safety hazards and risks are re-evaluated based on the metrics, with improvements to process safety systems proposed.

Phase 2.2 - Identify Common Metrics across Groups (Chapter 4)

The process safety performance metrics are re-affirmed or changed.

Phase 2.3 - Secure Leadership Support across Groups (Chapter 2)

Leadership recognizes the process safety systems gaps and approves the process safety system changes with clear and visible support for the improvements and re-affirmed or changed metrics.

Phase 3 – Act

Implement Changes to the SHEQ&S Program (Chapter 7)

The approved process safety system changes are implemented with the re-affirmed or changed metrics.

Phase 4 - Do

Implement the SHEQ&S Management Systems (Chapter 5)

Everyone uses operational discipline to apply the approved process safety system changes and improves the organization's conduct of operations at all levels.

These continuous improvement phases are not active all the time. Periodic reviews and visible leadership support for scheduling these process safety performance indicator reviews and addressing the detected performance gaps are critical for safe and reliable operations. A company's survival depends on these reviews.

3.6   EXAMPLES OF PERFORMANCE EFFECTS ACROSS SHEQ&S GROUPS

When the "big picture" process safety performance-related metric view is identified and monitored at the corporate level, less work is required across an organization since the closure of a process safety-related performance gap will close a gap that affected another SHEQ&S group. This section shows two cases on how process safety performance can positively and negatively affect the performance in other groups. Case 1 shows the overall operational risk improvement when the metrics are monitored (the positive result); Case 2 shows how lack of monitoring adversely affects the performance of the other groups (the negative result). Additional examples are provided in Chapter 8.

Case 1 (positive effect): Dow Chemical Company

A positive improvement in an organization's overall operational risk is shown with the improvement in Dow Chemical Company's process safety performance over a decade of tracking of process safety metrics [Overton 2008]. This is shown in Figure 3-3, with the 71% reduction of its process safety critical incidents (PSCM) corresponding with their loss of primary containment (LOPC) incidents, down 72%, and their injury and illnesses (down 84%) over the same time period.

From the report:

> "To put it another way, during the past 10 years, 13,000 employees did not suffer an injury or illness, 10,500 LOPCs did not occur, and 1,100 Process Safety incidents did not occur."

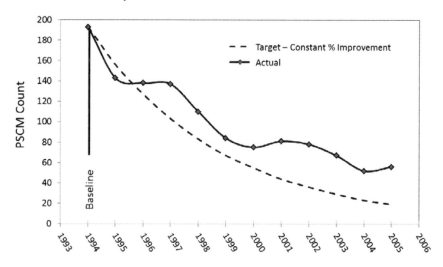

Figure 3-3. The improvement of Dow Chemical Company's process safety performance when monitoring and responding to process safety metrics

Case 2 (negative effect): Loss of containment incidents

The loss of containment of a hazardous material can have negligible or have significant adverse effects on the other groups, as is shown with the incidents in Table 3-2. The small phosgene release in West Virginia and the large methyl isocyanate (MIC) release at Bhopal caused fatalities [US CSB 2011b, Atherton 2008]. The small phosgene loss of containment affected the occupational safety and health group; the large methyl isocyanate loss of containment affected the occupational safety and health, environmental, quality and security groups (see qualifying notes that follow). In addition, the toxic release in Bhopal killed and injured thousands of people and contaminated the ground water.

Fast forward to today's world, using today's security-driven vulnerability assessments, a facility siting study would have identified the facility location in Bhopal as a high-risk facility due the quantity of MIC in storage and its proximity to the neighboring community. The fact that Union Carbide no longer exists, due in part to the incident in Bhopal, can be the same consequence that a company can have if its products do not meet the quality expectations of its customers. Hence, a leading indicator for ensuring quality products could be measuring the deviations from the set processing design conditions, the deviations triggering emergency responses and controlled process shutdowns. Had an effective SHEQ&S program been in place, these incidents may have been prevented with information shared across the organization that their operational risk was not being properly managed.

**Table 3-2. Adverse effects of loss of containment incidents
on the different SHEQ&S groups**

SHEQ&S Group		Small Release of phosgene (1)		Large Release of methyl isocyanate (2)	
		Leading Indicators (3)	Lagging Indicators (4)	Leading Indicators (3)	Lagging Indicators (4)
S	Process safety	Hazards and risk analysis; Equipment design and integrity; Operating limits	Fatality	Hazards and risk analysis; Equipment design and integrity; Operating limits; Emergency response	Fatalities
H	Occupational safety		Fatality		Fatalities
	Occupational health		Fatality		Fatalities
E	Environmental				Ground water contamination
Q	Quality				Business destroyed
S	Security			Vulnerability assessment	Community (public) fatalities
Number of groups with metrics which may affect process safety performance		5	3	6	6

Note

1	Reference: The US Chemical Safety Board (CSB), "Investigation Report, E.I. DuPont De Nemours & Co., Inc., Belle, West Virginia," Report No. 2010-6-I-WV, September 2011.
2	Reference: Atherton, John and Frederic Gil, "Incidents That Define Process Safety," CCPS/AIChE and John Wiley & Sons, Inc., Hoboken, NJ, 2008.
3	Not measured at the time of the incident; potential warning signs which could have been measured and tracked, helping prevent the fatalities across many of the process safety systems noted (e.g., equipment integrity tests and inspections could have prevented loss of containment with knowledge that equipment no longer met its design specifications).
4	Consequences of actual incident.

4 IDENTIFY COMMON METRICS ACROSS GROUPS

Although there are many metrics that can be selected to monitor and track process safety performance, care must be taken to ensure that the metrics chosen apply at each level in the organization. The corporate, facility and process unit level personnel responsible for improving process safety performance must be able to effectively track, monitor and respond to the metrics. Poorly selected metrics will hinder effective decision making.

This chapter provides an approach to identify metrics that affect process safety, helping integrate the metrics common between the SHEQ&S groups and helping to reduce the workload across the groups. The structure of this chapter is presented in Figure 4-1; an organization uses its existing metric tracking management system(s) to form the foundation for an effective SHEQ&S program. The key stakeholder values and external requirements, as well as any informal management systems, are described to help develop the SHEQ&S program in Sections 4.2 through 4.4. The SHEQ&S metrics selection team chartered to identify overlapping metrics is described in Section 4.5, with the team's metric prioritization approach presented in Section 4.6. This chapter continues with description of approaches to develop an initial, small scale pilot for the SHEQ&S program integrating the management systems for a baseline study (Section 4.7). Then the results from the baseline study are used to improve the SHEQ&S program before its full implementation across the organization (Section 4.8). Details on developing, implementing and responding to the pilot program are provided in more detail in Chapter 5, Section 5.3.

4.1 THE NEED FOR IDENTIFYING COMMON METRICS

Identifying common metrics is essential for successfully improving process safety performance. However, before starting on the design of an effective SHEQ&S program, it is important to understand how each group manages its group-specific metrics. S uccessfully reducing the work demands on the different SHEQ&S groups depends on understanding and enhancing the existing management systems, not creating new work processes. Fortunately, these management system differences and similarities can be identified at the same time that the metrics are being identified. Management systems exist at the process unit level, at the facility level, and at the corporate level. Once identified for each SHEQ&S group, similarities between these management systems can be evaluated and integrated, as needed, to reduce the monitoring work load. With this approach, it is important to recognize that the metrics selected for other levels are based on those selected at the process unit level, where the hazardous materials and energies are managed by operations personnel [CCPS 2010, CCPS 2011, and HSE 2006].

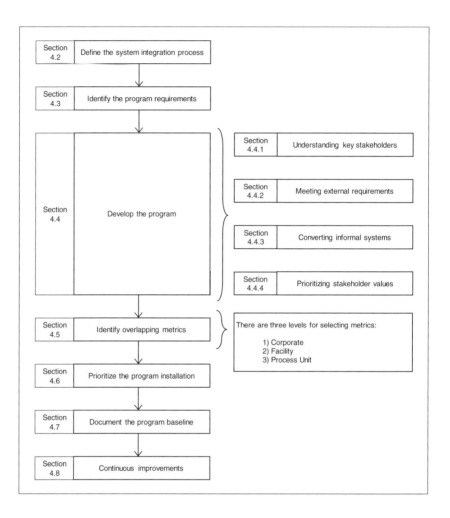

**Figure 4-1. The structure for Chapter 4:
Identifying common metrics across SHEQ&S groups**

4.2   DEFINE THE SYSTEM INTEGRATION PROCESS

The integration process must be defined with a cross-disciplinary team representing each of the SHEQ&S groups. This team should consolidate and agree upon the common risk areas, the common risk controls in place, the metrics that affect process safety performance, and the metrics that can be used to help each group improve its performance. When common risks are identified, the metrics chosen by the team should include leading and lagging indicators that will help each SHEQ&S group effectively monitor its risks. B y design of the integrated system, the process safety group will have the most metrics to monitor, as the focus of the SHEQ&S program is to identify metrics that affect process safety performance relative to the metrics monitored across all of the other groups.

4.3   IDENTIFY THE PROGRAM REQUIREMENTS

The SHEQ&S program requirements for monitoring the metrics needs to be clearly identified and maintained across all functional groups for the work process to effectively reduce the work load across the groups. B y understanding and enhancing existing management systems, synergies can be developed and reporting can be streamlined across and through the different organizational levels. The elements in the system should include clear definitions of the management and operational responsibilities at the corporate, facility, and process unit levels. Responsibilities for controlling and verifying raw materials and for production conditions and product quality should be clearly identified and appropriate for the level of the group. T he system should have traceability, inspection, and testing capabilities and there must be control of the measuring and testing equipment. There should be a defined process for identifying and addressing nonconforming results with clear steps on subsequent corrective actions. S ince these are some of the basic elements of a quality system, it makes sense to align the SHEQ&S program with those inherent to the Quality program [ISO 2008a].

4.4   DEVELOP THE PROGRAM

The SHEQ&S program is developed through an iterative process, with the team acknowledging the spirit of continuous improvement activities across all groups at every level in the organization. T he integrated system should align with and be consistent with a co mpany's culture and existing management style. H owever, since each company is different and, more importantly, each company's hazards and risks are different, this Guideline provides performance-based criteria only. Each company's program integration team should develop their own step-by-step prescriptive procedure, as needed.

In addition, there are group-specific requirements from external agencies and their key stakeholders which must be addressed by the program integration team. These SHEQ&S program expectations are shown in Figure 4-2.

4.4.1   Understanding the Key Stakeholders

The stakeholders are individuals or organizations that can (or believe they can) be affected by a process unit's operation, or who are involved with assisting or monitoring a facility's operation [CCPS 2010]. The key stakeholder values for each SHEQ&S group should be clearly communicated through both the metrics selection team and the program integration team. These values drive the group's performance metrics. Some examples of key stakeholders include:

- The process unit employees working with the hazardous materials and energies, who expect to leave the work place healthy;

- The people in the surrounding communities, who expect to live in a safe, healthy neighborhood;

- The owners, who expect to add value to the incoming raw materials and make profitable, quality products; and

- The regulators and community leaders, who expect that the operational risks have been reduced to protect the safety, health, and welfare of people and the environment.

4.4.2   Meeting external requirements

There are many external requirements that the members of the metrics selection team need to consider and incorporate during the process safety metric selection process. Regulatory requirements for reducing process safety-related risks, such as COMAH, the Seveso II Directive and the U.S. OSHA PSM and EPA RMP standards, are often tracked by covered companies with pre-defined process safety-related metrics. C ompanies that have ISO-related certifications, have actively joined the Responsible Care® program, or have networks through various process safety-related groups may have developed additional metrics. O ther external groups may have regulatory or certification-based requirements, as well. S ome of the external regulation and industry-related organizations with external requirements that need to be considered when the metrics selection team identifies and selects the company's metrics are listed in Appendix A.

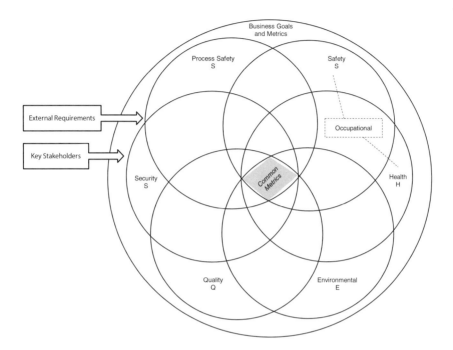

Figure 4-2. Requirements considered when developing the SHEQ&S program

4.4.3   Converting informal systems

Although the discussion up to this point has focused on identifying the existing management systems that lend themselves to the SHEQ&S program, there may be informal management systems currently in use that should be considered, as well. An informal system may exist if there is no good formal management system to address some program elements. An informal system exists, in part, because it works. Fortunately, these informal management systems can be identified at the same time that the metrics are being identified. Once identified, the informal system can be incorporated into the SHEQ&S program.

4.4.4   Prioritizing Stakeholder Values

There are many metrics that may be managed by other SHEQ&S areas but also affect process safety. For example, a significant process safety risk could occur with unauthorized access to a facility, issues which must be addressed by the security group. Security group metrics, such as "number of unauthorized access incidents" or "number of patrols completed on schedule," will help contribute to effectively managing the company's process safety risk. Hence, it is important to filter and prioritize all the metrics from each group. A stakeholder valuing and prioritization approach, based on process safety risk, is described in more detail in Section 4.5 below.

4.5   IDENTIFY OVERLAPPING METRICS

> *You cannot improve what you do not measure.*
> *-CCPS 2011*

Overlapping metrics between SHEQ&S groups that affect process safety performance can be identified and selected using a risk based process safety approach. This procedure is modeled after the CCPS Risk Based Process Safety guidelines, which are designed to help organizations handling hazardous materials and energies identify and address gaps in their process safety elements [CCPS 2007a]. Since harmful exposures to people and the environment can also occur when we lose containment of hazardous materials or energies, the risk-based approach helps us prioritize all SHEQ&S risks.

Since there are many steps for identifying and selecting appropriate metrics that affect process safety performance for the SHEQ&S program, the process map in Figure 4-3 has been created to help guide the metrics selection team. To help orient the metrics team on the main purpose for process safety, the scope can begin with a visualization of the incident "trajectory," starting at the equipment that is supposed to contain the hazardous material or energy. Process safety's goal is to keep hazards within the equipment and piping. There are three major safety management systems based on the organization's structure designed to manage the hazards: the process unit level, the facility level, and the company level (or process specific, then site level, then enterprise level [OECD 2008]). Once a process safety-related scenario is identified at a process unit, there are four sets of questions that screen for process safety-related outcomes across the SHEQ&S groups, which then help prioritize the metrics selection based on the outcome's risk to each group [CCPS 2007a].

These questions can be mapped through the organization at each level using some of the management assessment tools presented in Section 4.9. The management assessment tools guide the metrics team to the indicators being measured and to how they are analyzed. Resources used to measure the indicators, track them, and decide how to respond are identified, as well.

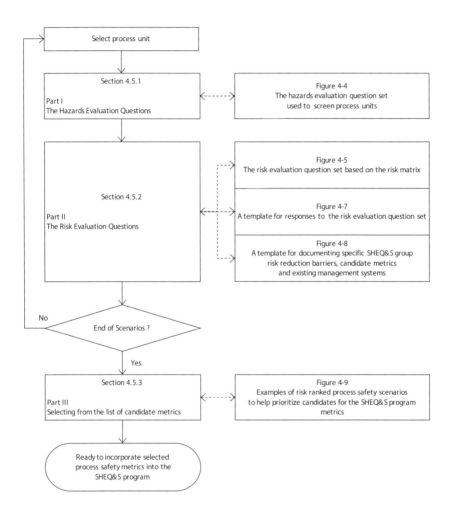

Figure 4-3. The process map for identifying metrics which affect process safety performance across groups

4.5.1   Part I – The Hazards Evaluation Question Set

There are four questions used to determine candidate metrics for the SHEQ&S program. T he first two, described in this section as "hazards evaluation" questions, help the metrics selection team screen for specific process units that may have metrics which affect process safety performance. The first question is process unit-specific, where operations manages and controls any hazardous material and energy:

Hazards Evaluation Question 1:   Does the process unit contain any hazardous materials or energies, including, but not limited, to those listed below?

Hazards of Materials:

Toxic, flammable, explosive, reactive, corrosive or unstable materials?

Process Design:

Temperature or pressure extremes, large inventories, etc.?

If there are no process safety-related hazards identified, then there are no metrics for the process unit (the answer to Question 1 is "No") and the process unit does not have any process safety-related hazards. I t can be eliminated from the integration effort. If the answer to Question 1 is "Yes," then the metrics team proceeds to Question 2.

The second hazards evaluation question is SHEQ&S group-specific:

Hazards Evaluation Question 2:   Can the process safety-related hazard cause any of these consequences?

S	Process Safety	Harm to people on-site at the facility, harm to people off-site in the surrounding communities (i.e., fatalities; both irreversible and reversible injuries)
S	Occupational Safety	Harm to people on-site at the facility (i.e., fatalities; both irreversible and reversible injuries)
H	Health	Harm to people on-site at the facility (i.e., acute; chronic; irritation)
E	Environmental	Harm to the environment (i.e., air, land and water; contamination)
Q	Quality	Harm to the customer and to the business (i.e., product fails to meet specifications or injures the consumer; product stewardship)
S	Security	Harm to people off-site, in the surrounding communities or by use of the materials elsewhere (i.e., a result of terrorist activity)

If the answer to any SHEQ&S group from Question 2 is "No," there are no consequences that affect process safety performance for that particular group. The conclusion: Risk Level A [CCPS 2007a]. The particular SHEQ&S group does not have a metric in this scenario that is a part of the integrated effort.

If the answer to any SHEQ&S group from Question 2 is "Yes," then the metrics selection team proceeds to the risk evaluation questions in Question Set 3 and 4. The hazards evaluation questions (Questions 1 and 2) and the metrics selection team's responses are summarized in Figure 4-4.

4.5.2   Part II – The Risk Evaluation Question Set

Once the metrics selection team determines from the two "hazards evaluation" questions that there are process safety-related hazards with outcomes affecting process safety performance, the team addresses each of the affected groups with a series of risk evaluation questions to help differentiate between potentially high risk process safety scenarios, such as those with multiple fatalities, and potentially low risk events, such as those causing temporary health effects. The metrics selection team's first three risk evaluation questions focus on the risk matrix and are combined into Question Set 3 as follows:

Risk Evaluation Question Set 3:

Question 3.1: "What can go wrong?"	(Assume failure of all barriers; "worst case" outcome)
Question 3.2: "How bad could it be?"	(Outcome's severity; its consequence)
Question 3.3: "How often might it happen?"	(Event's likelihood; its frequency)

The answers to Questions 3.1, 3.2 and 3.3 should be based on the hazards associated with the process unit, and then compared to a company's tolerable risk matrix was shown generically in Figure 2-3. For the purposes of identifying candidate metrics at this point, the metrics selection team should simplify their corporate risk matrix (sometimes 3x3, 4x4, 5x5 or 6x6 matrices) and focus on the extremes shown in a 2x2 matrix, using the RBPS approach with either a High (H) or a Low (L) option, only. The resulting frequency times the consequence risk ranking is simply HH, HL, LH or LL, as is shown with the 2x2 risk matrix and questions in Figure 4-5.

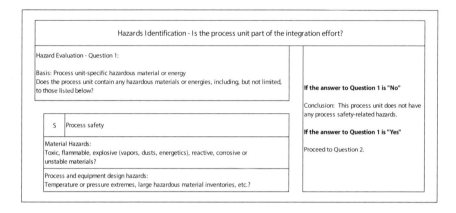

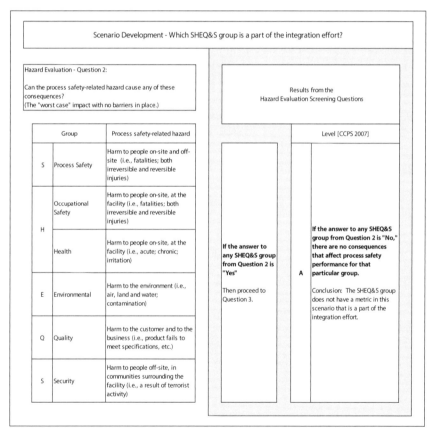

**Figure 4-4. The hazards evaluation question set
used to screen for process units**

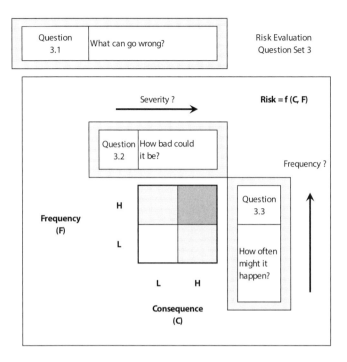

Figure 4-5. The risk evaluation question set based on the risk matrix

The risk levels defined with the RBPS approach are used when the metrics selection team answers the risk-based questions 3.1, 3.2 and 3.3, to help prioritize the scenarios from which potential candidate metrics can be selected. For reference, these RBPS risk levels are shown in Figure 4-6, ranging from a "Level E" (with the high consequence, high frequency, high risk event) to a "Level B" (the low consequence, low frequency, low risk event) [CCPS 2007a, Chapter 23, Figure 23.1]. Note that the "Level A" risk level for the identical CCPS questions in Figure 4-6 is the response *eliminated* from further evaluation with Hazard Evaluation Question 2 above. The metrics selection team's approach for Question Set 3 is shown in Figure 4-7.

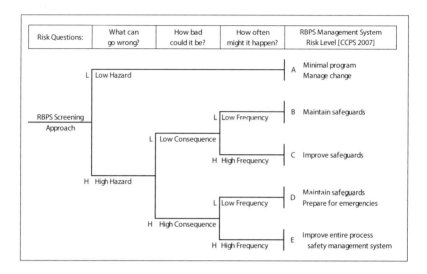

**Figure 4-6. The CCPS Risk Based Approach
(for reference to the approach shown in Figure 4-7)**

The results for selecting potential candidates for each SHEQ&S group from this risk based metric candidate assessment are as follows:

Risk Level E	"Select barrier-related metrics"	(First choice for metric candidates)
Risk Level D	"If no options in E, Consider barrier-related metrics"	(Second choice)
Risk Level C	"If no options in D or E, Consider barrier-related metrics"	(Third choice)
Risk Level B	"If no options in C, D, or E, Consider barrier-related metrics"	(Last choice)
Risk Level A	Since this risk level was determined at Risk Question 2 (see Figure 4-4), there are no process safety risks identified for the process unit.	

The metrics selection team's next risk evaluation question, Question 4, is also shown in Figure 4-7:

Risk Evaluation Question 4: " What barriers currently exist, need upgrading, or need to be in place to reduce the process safety risk?"

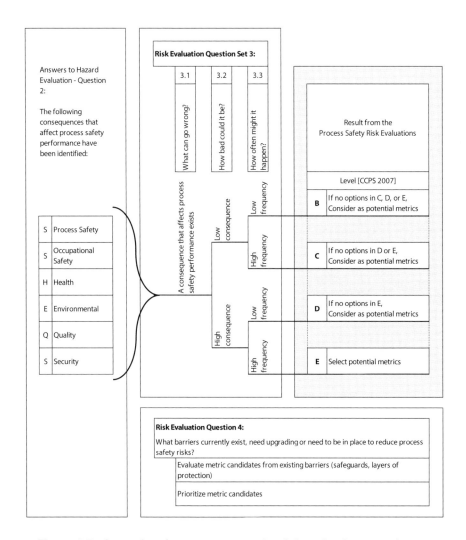

Figure 4-7. A template for responses to the risk evaluation question set

At this point, the work of the process safety group can be used to identify these barriers, which are defined as the "layers of protection" that have been implemented to reduce the process safety risk. Using the visual Bow Tie tool described in Chapter 3 (Figure 3-1), the metrics selection team can compare and discuss the potential metrics for each group. Each SHEQ&S group member documents their group's consequences based on the loss of containment scenario being reviewed, and with the diagram in Figure 4-8, the team members from the affected group can document what barriers they take into account, as well. There are barriers and controls in place to ensure that no gaps (the "holes" in the Swiss cheese model) exist, or if gaps do exist, then actions can be taken to fill the gaps. These preventive and mitigative barriers are designed to reduce the event's risk. By tracking and monitoring them in an integrated system, the organization becomes more effective in managing its process safety risk and can show measurable improvements in its process safety performance.

Also included in Figure 4-8 are rows and columns for documenting both the formal and informal SHEQ&S systems that are currently being used to track and manage these metrics. As was discussed earlier, the foundation has been set for comparing, prioritizing and choosing key metrics for the SHEQ&S program now that the similarities and differences between the metrics and their tracking systems have been identified.

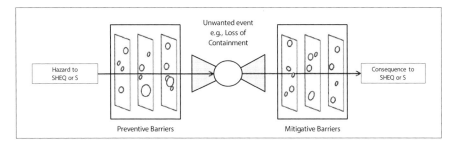

Process Safety Barriers and Candidates for the Process Safety Metrics

	Group	Current Metrics being monitored and tracked; Current preventive barriers in place	Current metrics being monitored and tracked; Current mitigative barriers in place
S	Process safety		
S	Occupational Safety		
H	Health		
E	Environmental		
Q	Quality		
S	Security		

Metric Tracking and Monitoring Systems (Formal and Informal)

	Group	System(s) for the preventive barriers	System(s) for the mitigative barriers
S	Process safety		
S	Occupational Safety		
H	Health		
E	Environmental		
Q	Quality		
S	Security		

Figure 4-8. A template for documenting specific SHEQ&S group risk reduction barriers, candidate metrics and existing management systems

4.5.3 Part III – Selecting from the list of candidate metrics

From the risk based analysis described in Section 4.5.2, the identified metrics and their associated barriers become the basis for determining both potential leading and lagging metrics for the SHEQ&S program.

The results from the discussions in this guideline up to this point are:

- Recognizing that there are potentially overlapping metrics that affect process safety performance (see Figure 1-2) – Chapter 1
- Securing leadership support across the groups for integrating the SHEQ&S management systems – Chapter 2
- Clearly addressing each group's risks (see Figure 3-1) – Chapter 3
- Identifying candidates for the metrics which affect process safety performance (see Figure 4-8) – Chapter 4
- Identifying the tracking systems that can be integrated (see Figure 4-8) – Chapter 4

Each scenario will have a series of potential metrics, based on the results using the questions in Figure 4-7 and listed in Figure 4-8. An example using different potential events is shown in Figure 4-9, from which the integrating team can prioritize each event's risk-ranked process safety metric.

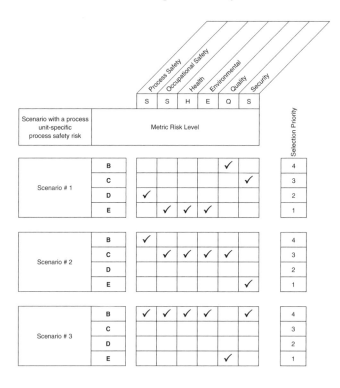

Figure 4-9. Examples of risk ranked process safety scenarios to help prioritize candidates for the SHEQ&S program metrics

Selecting from these process safety metric candidates should be easier now that they have been risk ranked and are essentially "prioritized" across each SHEQ&S group. The prioritization method for selecting metrics which affect process safety performance has, to some extent, been identified using the company's tolerable risk matrix. S ince the metrics selection team used the risk based approach to identify candidates from the beginning of their review, the discussions for the key metrics should focus at this point on selecting from the candidate at risk level "E" first, then from "D" through "B" (recognizing that these levels may differ in importance with respect to each SHEQ&S group. An illustration using the Bhopal event shows how to use Figure 4-4 through Figure 4-9 (from Sections 4.5.1 through 4.5.3 for selecting potential metrics) and is described below in Section 4.5.4 (refer to the chapter framework shown in Figure 4-3). The process for implementing the SHEQ&S program based on these metrics will be described in the subsequent chapters.

4.5.4   An example for how to identify overlapping metrics

This section illustrates how the metrics selection team can use the approach introduced in Section 4.5.1 through Section 4.5.3, using the incident that occurred in Bhopal in 1984. Th is watershed process safety incident changed the way process safety was managed, including efforts to incorporate process safety principles in the chemical engineering curriculum, the formation of industry-wide process safety risk reduction efforts, the formation of the CCPS (1985), and the promulgation of the U.S. OSHA Process Safety Management (PSM) standard eight years later in 1992. D ecades after the event, the water in the neighborhoods around the Bhopal site is still contaminated, with both survivors and current area residents continuing to suffer from adverse health and environmental effects [Willey 2007].

The goal of the exercise is to identify metrics that affect process safety performance across the different SHEQ&S groups using a p rocess safety risk based approach. Once the risk based metrics have been identified and selected, the following chapters can be used for reference as to how an integrated management SHEQ&S system can be designed and implemented in an organization. T he process for identifying additional metrics from each scenario is iterative, with the steps noted in Sections 4.5.4.1 through 4.5.4.4 repeated for each scenario. Prioritizing and selecting from the candidate metrics list becomes the team's next task, based on the same risk-based approach as is described in Section 4.5.4.4 below.

4.5.4.1   Brief description of the incident at Bhopal

The toxic release that occurred in Bhopal in 1984 killed thousands and injured hundreds of thousands of people [Kletz 2009]. T he initiating event was water reacting with the contents of an intermediate storage tank to produce a h eavier-than-air toxic gas, methyl Isocyanate (MIC). P ressure relief equipment was activated (the "loss of containment" from the storage tank), but the caustic scrubber, the process flare, and the water spray systems designed to neutralize and contain the escaping MIC failed such that the MIC passed through all of the mitigative barriers. The toxic MIC gas drifted over the fence line with disastrous impact on the surrounding community. The emergency responders did not know how to handle the release or how to manage the hazardous event, resulting in thousands of deaths to people living in the surrounding communities. This incident devastated Union Carbide, a company which was incorporated in 1917 and developed an economical way to make ethylene from natural gas in 1920 which gave birth to the modern petrochemicals industry. Even with almost eight decades of chemical industry experience, the tragedy at Bhopal still occurred. Union Carbide built its history on what was at the time an industry benchmark using a thoroughly-ingrained "Safety First" culture to manage its hazardous chemicals and processes. "It was a deeply ingrained commitment that involved every employee worldwide and had been spurred in the chemical business by stringent internal standards dating back to the 1930s." [Browning 1993]

4.5.4.2   Part I – The Hazards Evaluation Questions

The hazards evaluation questions were introduced in Figure 4-4 and are populated in this example with the answers generated by the metrics selection team members. As is listed in Figure 4-10, the answer is "Yes" to the first hazards evaluation question: there is potential for a significant toxic release.

Based on this result, the metrics selection team proceeds to the second hazards evaluation question, as is shown in Figure 4-11, with the answer "Yes" to every SHEQ&S group. Now the metrics selection team has established consensus with all members that a process safety-related consequence exists for each group. A potentially adverse effect to their specific group's performance could occur if all the preventive and mitigative barriers designed to reduce the process safety risk fail (the "Swiss Cheese" effect). With the results from Question 2 answered, the metrics selection team is ready to proceed with the risk evaluation questions in Part II of the metric selection process.

Hazards Identification - Is the process unit part of the integration effort?	
Hazard Evaluation - Question 1: Basis: Process unit-specific hazardous material or energy Does the process unit contain any hazardous materials or energies, including, but not limited, to those listed below?	
	If the answer to Question 1 is "No"
S	Process safety
Material Hazards: Toxic, flammable, explosive (vapors, dusts, energetics), reactive, or unstable materials?	Conclusion: This process unit does not have any process safety-related hazards.
Answer for Bhopal Example: Yes **Vessel contents are water reactive; runaway reaction will produce toxic MIC vapors**	**If the answer to Question 1 is "Yes"**
Process and equipment design hazards: Temperature or pressure extremes, large hazardous material inventories, etc.?	Proceed to Question 2.
Answer for Bhopal Example: Yes **Vessel inventory is large**	

Figure 4-10. Answers to the first hazards evaluation question

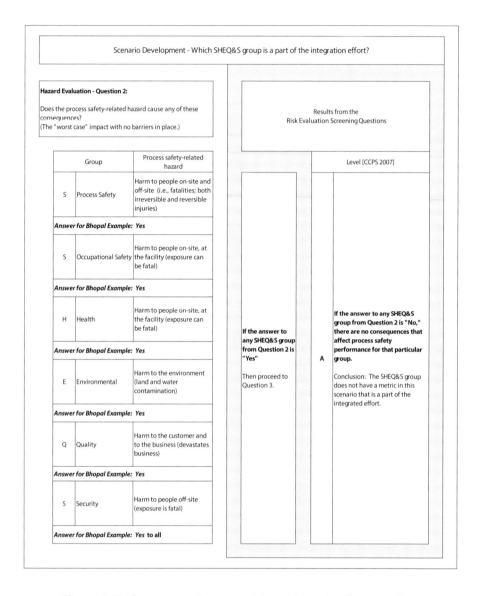

Figure 4-11. Answers to the second hazards evaluation question

4.5.4.3   Part II – The Risk Evaluation Questions

The answers to the hazards evaluation questions in Part I have directed the metrics selection team to the next step: answering the risk evaluation questions that were shown in Figure 4-5 and Figure 4-7, with documentation of the specific barrier metrics and systems in place shown in Figure 4-8. Using the general risk matrix shown in Figure 4-5, the metrics selection team continues with each question, noting the answers to the risk evaluation questions for scenario development based on the risk ranking for each SHEQ&S group in Figure 4-12. (Note assumptions in Figure 4-12 are for illustration purposes.)

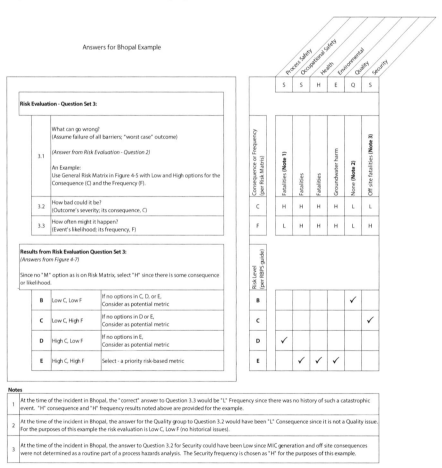

Figure 4-12. Answers to the risk evaluation question set

The metrics selection team's next question is Risk Evaluation Question 4: "What barriers currently exist, need upgrading, or need to be in place to reduce the process safety risk?" with the answers listed in the template shown in Figure 4-13. The risk reduction efforts from the process safety group can be used to populate the template in Figure 4-13, as well, using hazards and risk analyses which identify the barriers and layers of protection [CCPS 2001, CCPS 2009a]. These preventive and mitigative barriers are identified to help reduce process safety risk. With the metrics identified, the metrics selection team discusses and then populates the template in Figure 4-14 with the existing SHEQ&S management systems currently tracking the identified metrics. The answers documented by each SHEQ&S group in Figure 4-13 and Figure 4-14 set the stage for comparing, prioritizing and choosing the key metrics for the SHEQ&S program, as discussed in the next section.

4.5.4.4   Part III – Selecting from the list of metric candidates

The metrics chosen for the SHEQ&S program can be selected from the list of metrics created in Figure 4-13. The process unit metrics list may be long, due to a complex process safety system designed to manage the highly hazardous materials. However, remember that this long list becomes the foundation for the shorter, aggregated lists being monitored at the facility and the corporate levels [HSE 2006, OECD 2008, CCPS 2010, and CCPS 2011]. In other words, the list of metrics needed for tracking and monitoring will be shorter for higher levels in the organization, providing a more effective process safety monitoring within the SHEQ&S program. This is particularly true since the metrics being monitored must be ones that can be influenced by the decisions of the group monitoring them.

As is shown in Figure 4-15, the process safety metric priority selection criteria is based on the metric's risk level, with the most significant metrics corresponding to the high risk scenarios. For the purposes of this exercise, a prioritized "order" would begin with the metrics identified in the occupational safety and health and environmental groups (risk level E), the process safety group (risk level D), the security group (risk level C) and then the quality group (risk level B). It is a double-edged sword when there are too many metrics identified and from which to choose: there are many to select from (a good thing), but which to choose (the need to prioritize). When there are too many to choose from, there may be some "low hanging fruit" which are easy to reach and which can show quick measurable process safety performance improvements. For discussion on this approach, refer to Chapter 5, Section 5.3, piloting the SHEQ&S program.

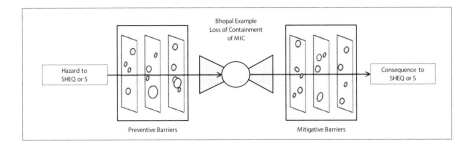

Bhopal Example: Process safety barriers and scenario metric candidates which affect process safety performance

Affected group, based on results from screening questions		"Example" Preventive Barriers in place and Metrics being monitored and tracked	"Example" Mitigative Barriers in place and Metrics being monitored and tracked
S	Process safety	Process design (redundant and refrigerated vessels; instrumentation and alarms) Metrics:	Equipment design (pressure relief; instrumentation and alarms) Metrics:
		Maintenance procedures Metrics:	Emergency equipment design (instrumentation and alarms; caustic scrubber; process flare; water spray) Metrics:
		Operating procedures and training Metrics:	Community alarms Metrics:
S	Occupational Safety	Procedures and Training Metrics:	Emergency response Metrics:
H	Health	Procedures and Training Metrics:	Metrics:
E	Environmental	Process design (redundant and refrigerated vessels; instrumentation and alarms) Metrics:	Emergency equipment design (instrumentation and alarms; caustic scrubber; process flare; water spray) Metrics:
		Maintenance procedures Metrics:	Metrics:
		Operating procedures and training Metrics:	Metrics:
Q	Quality	Metrics:	Metrics:
S	Security	Facility siting studies Metrics:	Metrics:

Figure 4-13. An answer template for the risk evaluation question set

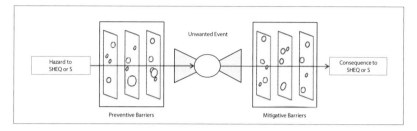

Metric Tracking and Monitoring Systems (Formal and Informal)		
Group with formal and/or informal metric tracking system	Preventive Barrier Tracking System(s)	Mitigative Barrier Tracking System(s)
S	Process safety		
S	Occupational Safety		
H	Health		
E	Environmental		
Q	Quality		
S	Security		

Figure 4-14. An answer template for documenting existing SHEQ&S management systems

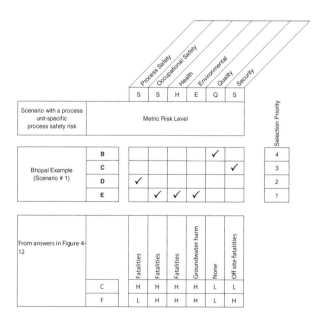

Figure 4-15. The process safety metric priority selection criteria based on the metric risk level

To illustrate the metrics prioritization process using the metrics selected from the Bhopal incident, Figure 4-9 was used to produce Figure 4-16. Based on these qualitative risk ranked metrics results, the metrics monitoring the layer or layers of protection preventing or mitigating the generation of MIC in the storage tank impact the process safety, the occupational safety and health and the environmental groups the most, with risk levels D and E. Due to the magnitude of human suffering and environmental impact of the release at Bhopal, many have pointed to the types of failures that, with hindsight, make selecting barrier-related process safety indicators somewhat obvious. However before 1984, identification of specific process safety elements as well as the integrated process safety management system did not exist as we know them today [Vaughen 2012].

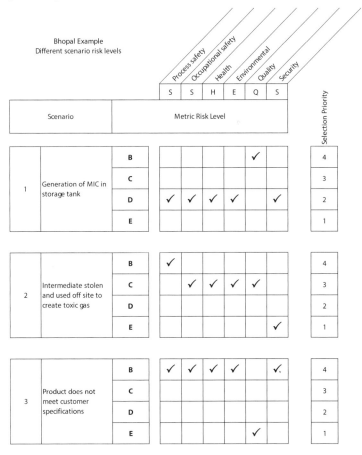

Figure 4-16. The process safety metric priority selection criteria based on the metric risk level

As can be seen in the example Figure 4-16, the loss of containment of a hazardous material adversely affected the other SHEQ&S group's performance metrics. The toxic release killed and injured people (process safety and health) and contaminated the ground water (environmental). Using today's vulnerability assessments, a facility siting study would have identified the facility location as high-risk due to the quantity of MIC in storage (security). The safety and environmental consequences of the release eventually devastated the company, the same result that would have occurred if it had not been able to satisfy its customers with an acceptable (safe) product and subsequently lost its market (quality). Using common risk-based metrics will lead to more effective metric monitoring, tracking, and response efforts across the organization, resulting in improved process safety performance and company longevity.

4.6   PRIORITIZE THE PROGRAM INSTALLATION

Although there are no hard and fast rules for deciding the order in which to install a program, the likelihood for reducing the work load across the SHEQ&S groups will be limited without first developing a clearly defined program of integrated management systems. Without a program in place, the new and more effective tracking and monitoring procedures identified by the program integration team will most likely be ignored. The organization will carry on as before, delaying any potential improvements in process safety performance.

It is important to recognize at this point that the SHEQ&S program "installation" need not be a new system. Recall as the metrics selection team developed its answers to the risk based questions in Section 4.5, it identified and documented currently existing programs monitoring the SHEQ&S metrics. If program gaps are discovered in this review, specific SHEQ&S program design and development support will need to be secured across all levels of the organization to address and correct the deficiencies before installing the SHEQ&S program.

4.6.1   Identifying the process safety areas associated with each metric

The first step for prioritizing and selecting the metrics for the integrated system is to identify which process safety area applies to the metric that has been identified. In general, the ten fundamental areas under which an effective, integrated process safety and risk management program operates are as follows:

1)   Overall process safety management ("PSM") framework

2)   Process technology (i.e., hazards of materials, process design, equipment design)

3)   Process hazards analyses (including layers of protection, human factors and/or facility siting analyses)

4) Process operations (including procedures and safe work practices for all personnel)

5) Process maintenance (including equipment integrity and maintenance-specific procedures)

6) Emergency response (includes internal as well as external resources for the facility)

7) Incident investigations (includes facility recovery efforts and sharing and communicating findings)

8) Managing changes (includes safe handovers to operations and maintenance, once a change is made)

9) Monitoring process safety system performance (includes compliance and/or systems audits, tracking leading indicators, etc.)

10) Organizational capability (including leadership, safety culture, operational discipline, conduct of operations, and training).

As is shown in Figure 4-17, these ten process safety areas can be integrated with the other SHEQ&S groups due to their inherent similarities with the other management systems.

CCPS has expanded the ten general process safety areas into twenty Risk Based Process Safety Elements, or "pillars," to help identify specific areas that need to be addressed for an effective process safety management program [CCPS 2007a, Sepeda 2010]. Hence, by adding these specific areas to the general areas noted in Figure 4-17, we obtain Figure 4-18. It is interesting to note in Figure 4-18, as well, that the foundation of any successful process safety management program is built with strong leadership, a strong process safety culture, effective operational discipline, effective conduct of operations and an integrated effort on the different management systems in each of the process safety areas. These foundations have been identified within the organization's capabilities for implementing a successful process safety and risk management program [CCPS 2007a, Klein 2015].

To help the metrics selection team identify which process safety area is being used to monitor the candidate metrics, Figure 4-19 can be used to summarize both preventive and mitigative process safety areas and Figure 4-20 can be used as a reference to the specific pillars using the CCPS RBPS elements.

Process Safety Area	Process safety (S)	Occupational safety (S)	Health (H)	Environmental (E)	Quality (Q)	Security (S)
1 — Overall process safety management ("PSM")	✓					
2 — Process technology (i.e., hazards of materials, process design, equipment design)	✓	✓	✓	✓	✓	
3 — Process hazards analyses (including layers of protection, human factors and/or facility siting analyses)	✓			✓		✓
4 — Process operations (including procedures and safe work practices for all personnel)	✓	✓	✓	✓	✓	
5 — Process maintenance (including equipment integrity and maintenance-specific procedures)	✓	✓	✓	✓	✓	
6 — Emergency response (includes facility-specific as well as surrounding community resources)	✓	✓	✓	✓		✓
7 — Incident investigations (includes securing evidence, facility recovery efforts and sharing/communicating findings)	✓	✓	✓	✓	✓	✓
8 — Managing changes (includes safe handovers to operations/maintenance once change is made)	✓	✓	✓	✓	✓	✓
9 — Monitoring process safety system performance (includes, compliance and/or systems audits, tracking leading indicators, etc.)	✓	✓	✓	✓	✓	✓
10 — Organizational capability (including leadership, safety culture, operational discipline, conduct of operations and training)	✓	✓	✓	✓	✓	✓

Figure 4-17. The common process safety areas mapped across each SHEQ&S group

Process Safety Area / CCPS Risk Based Process Safety Chapter Number (2007)			Process safety	Occupational safety	Health	Environmental	Quality	Security
1	Overall process safety management ("PSM")		S	S	H	E	Q	S
2	Process technology (i.e., hazards of materials, process design, equipment design)		✓	✓	✓	✓	✓	
	8	Process knowledge management						
3	Process hazards analyses (including layers of protection, human factors and/or facility siting analyses)		✓			✓		✓
	9	Hazard identification and risk analysis						
4	Process operations (including procedures and safe work practices for all personnel)		✓	✓	✓	✓	✓	
	10	Operating procedures						
	11	Safe work practices						
5	Process maintenance (including equipment integrity and maintenance-specific procedures)		✓	✓	✓	✓	✓	
	12	Asset integrity and reliability						
6	Emergency response (includes facility-specific as well as surrounding community resources)		✓	✓	✓	✓		✓
	18	Emergency management						
7	Incident investigations (includes securing evidence, facility recovery efforts and sharing/communicating findings)		✓	✓	✓	✓	✓	✓
	19	Incident investigation						
8	Managing changes (includes safe handovers to operations/maintenance once change is made)		✓	✓	✓	✓	✓	✓
	15	Management of change						
	16	Operational readiness						
	23	Implementation						
9	Monitoring process safety system performance (includes, compliance and/or systems audits, tracking leading indicators, etc.)		✓	✓	✓	✓	✓	✓
	4	Compliance with standards						
	20	Measurement and metrics						
	21	Auditing						
	22	Management review and continuous improvement						
10	Organizational capability (including leadership, safety culture, operational discipline, conduct of operations and training)		✓	✓	✓	✓	✓	✓
	3	Process safety culture						
	5	Process safety competency						
	6	Workforce involvement						
	7	Stakeholder outreach						
	13	Contractor management						
	14	Training and performance assurance						
	17	Conduct of operations						

Figure 4-18. The common process safety areas mapped across each SHEQ&S group with reference to the CCPS Risk Based Process Safety Elements (RBPS)

Process Safety Area	Process Safety Metric Candidates for the SHEQ&S Program	
	Preventive Risk Reduction Effort	Mitigative Risk Reduction Effort
	Metric (s)	Metric (s)
1	Overall process safety management ("PSM")		
2	Process technology (i.e., hazards of materials, process design, equipment design)		
3	Process hazards analyses (including layers of protection, human factors and/or facility siting analyses)		
4	Process operations (including procedures and safe work practices for all personnel)		
5	Process maintenance (including equipment integrity and maintenance-specific procedures)		
6	Emergency response (includes facility-specific as well as surrounding community resources)		
7	Incident investigations (includes securing evidence, facility recovery efforts and sharing/communicating findings)		
8	Managing changes (includes safe handovers to operations/maintenance once change is made)		
9	Monitoring process safety system performance (includes, compliance and/or systems audits, tracking leading indicators, etc.)		
10	Organizational capability (including leadership, safety culture, operational discipline, conduct of operations and training)		

Figure 4-19. Matrix for identifying common metrics across each of the process safety areas

Process Safety Area			Process Safety Metric Candidates for the SHEQ&S Program	
			Preventive Risk Reduction Effort	Mitigative Risk Reduction Effort
		CCPS Risk Based Process Safety Chapter Number (2007)	Metric (s)	Metric (s)
1		Overall process safety management ("PSM")		
2		Process technology (i.e., hazards of materials, process design, equipment design)		
	8	Process knowledge management		
3		Process hazards analyses (including layers of protection, human factors and/or facility siting analyses)		
	9	Hazard identification and risk analysis		
4		Process operations (including procedures and safe work practices for all personnel)		
	10	Operating procedures		
	11	Safe work practices		
5		Process maintenance (including equipment integrity and maintenance-specific procedures)		
	12	Asset integrity and reliability		
6		Emergency response (includes facility-specific as well as surrounding community resources)		
	18	Emergency management		
7		Incident investigations (includes securing evidence, facility recovery efforts and sharing/ communicating findings)		
	19	Incident investigation		
8		Managing changes (includes safe handovers to operations/maintenance once change is made)		
	15	Management of change		
	16	Operational readiness		
	23	Implementation		
9		Monitoring process safety system performance (includes, compliance and/or systems audits, tracking leading indicators, etc.)		
	4	Compliance with standards		
	20	Measurement and metrics		
	21	Auditing		
	22	Management review and continuous improvement		
10		Organizational capability (including leadership, safety culture, operational discipline, conduct of operations and training)		
	3	Process safety culture		
	5	Process safety competency		
	6	Workforce involvement		
	7	Stakeholder outreach		
	13	Contractor management		
	14	Training and performance assurance		
	17	Conduct of operations		

Figure 4-20. Matrix for identifying common metrics across each of the process safety areas with reference to the CCPS Risk Based Process Safety Elements (RBPS)

4.6.2   Some steps for the SHEQ&S program implementation

Based on the elements in a quality management system, the following order can be used to identify the steps for implementing the SHEQ&S program [ISO 2008a]:

1) Establish clear management and operational responsibility at the corporate, facility and process unit level.

2) Coordinate with purchasing the control and verification of raw materials handled in the hazardous processes.

3) Establish (through operations) the hazardous process production conditions.

4) Establish (through operations and maintenance) hazardous-duty equipment and piping traceability, inspection, and testing capabilities.

5) Define the process for identifying and addressing nonconforming results on the hazardous-duty equipment with clear, documented steps on subsequent corrective actions, if needed.

The first four steps frame the discussions in Chapter 2, where support at every level of the organization is essential for an effective SHEQ&S program. The fifth step is defined in Chapter 6, where the "Check" phase describes how the SHEQ&S performance is monitored. Chapter 7, the "Act" phase, describes how subsequent changes based on deficiencies can be implemented.

4.7   DOCUMENT THE PROGRAM BASELINE

The SHEQ&S program baseline is defined in this guideline as the current state of the existing metrics across the organization. The baseline provides a clearly defined starting point at the time the system implementation begins and from which performance improvements can be measured. Hence, it is important for each company to recognize that its "first pass" with the SHEQ&S program based on metrics which affect process safety performance may provide disconcerting results in the beginning, especially since some of the metrics may be new measures that were not identified and were not tracked in the past. In these situations everyone at all levels of the organization must have patience, recognizing that all new efforts take time to be established, and that it may be weeks, if not months, before observing measurable progress in the process safety performance.

4.8   CONTINUOUS IMPROVEMENTS

The SHEQ&S program is designed considering the continuous improvement life cycle: the Plan, Do, Check and Act phases shown in Chapter 1, Figure 1-3. The SHEQ&S program will evolve over time, becoming better with its implementation at all organizational levels as everyone applies their expertise and experience to identify deficiencies and make improvements. T he continuous improvement monitoring phases are proactive by design, using and responding to particular leading process safety indicators identified by the metrics selection team. The improvement phases to identify and address process safety system strengths and weaknesses and to clearly distinguish between compliance and performance-related gaps for the different process safety systems are described in more detail in Chapter 6.

It is hoped that this guideline will help a company evolve in a proactive manner, improving its process safety performance over time with level-appropriate selection, monitoring and implementations of its metrics developed as discussed in this chapter (selection) and in Chapters 6 and 7 (monitoring and implementation, respectively). Periodic SHEQ&S program reviews of these metrics will verify the current metrics, with the potential to discover additional gaps that will need to be addressed and then shared with others across the organization. T he goal is to prevent a company from having to respond to an emergency process safety event like the Bhopal incident, which caused tremendous human suffering and eventually devastated Union Carbide.

4.9   SOME MANAGEMENT SYSTEM ASSESSMENT TOOLS

This section provides additional information and details on some management system assessment tools available that can help a company design and implement an effective SHEQ&S program. A s was noted earlier in Section 4.4, the stakeholders who make decisions must understand how their decisions affect the overall operational risk. T he interactions are complex. The system assessment tools described in this section help define the scope, prioritize the resources essential to meet measurable objectives, define who is responsible and identify what training and knowledge these personnel will need. T hese measurable objectives have been prioritized through the risk based approach described in Section 4.5. The mapping tools help define the standard that must be achieved to measure success, describing what is being done, how it is being done, when it is being done, and who is doing it.

In general, "process mapping" is used to understand how management processes flow through and across an organization. The objectives of process mapping, the stream of activities used to transform a well-defined input or set of inputs into a pre-defined set of outputs, are depicted in a g eneral diagram for process mapping in Figure 4-21. The key process inputs and outputs are clearly identified. P rocess mapping is recommended when implementing quality management systems, as the effectiveness of the work processes improves when everyone's activities are clearly identified, implemented and managed [ISO 2008b]. P rocess mapping communicates the work processes and activities by showing the inputs, tasks, interfaces and outputs through visual maps and flowcharts. Hence, once the integrated SHEQ&S team has completed its process mapping assessment, there should be no uncertainty as to the requirements for each SHEQ&S group.

The assessment tools show the sequence of the decision-making process in the organization and how decisions are made. A detailed map typically depicts the process using rectangular boxes, flows with arrows between boxes, decision with diamonds, and options for a decision written on the lines exiting that decision. Since decisions at all levels of the organization will directly or indirectly influence how resources are allocated and how personnel will respond, the decisions have the potential to adversely affect the performance of another group. Personnel unfamiliar with the process should be able to read and understand the interactions that occur during the work flow process using one or more of these assessment tools. Depending on the level of detail, the tools may show the delays and other inefficiencies inherent in the existing work process, as well as providing the organization with its improvement opportunities. Analyzing process maps and flow charts will help streamline the SHEQ&S program and ensure that each SHEQ&S group clearly understands their role in helping manage process safety.

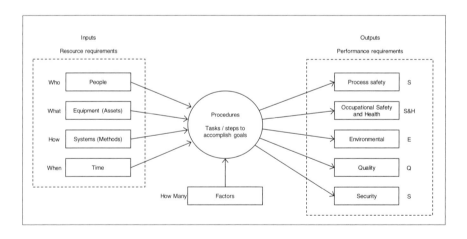

Figure 4-21. The objectives when mapping a process

As is shown in Figure 4-21, process mapping combines the elements of a work flow, selecting from personnel, materials, equipment, methods, and time information to show tasks and results. The human resources, asset resources, and system resources are defined. From the SHEQ&S program integration team's perspective, the process unit level contains the equipment required to control the process hazards (these could be called "assets" or "machines," too). The facility and corporate level process safety systems help the organization apply consistent procedures. The mapping tools make work visible to everyone, showing who is doing what, with whom, and depending on the detail, when and for how long the task is performed. If mapped against an organizational chart, process mapping helps identify inefficiencies, coverage gaps, and responsibility gaps. If mapped against the procedural steps and tasks, process mapping helps identify bottlenecks, duplications, unnecessary steps, sources of delay, rework (fixing errors instead of preventing them), cycle time, and responsibility ambiguities. There are three types of management assessment tools used in process mapping:

- Process maps or relationship maps
- Swim lane charts or cross-functional maps
- Process flow charts or process flow diagrams

Each of these assessment tools is described in more detail below, in the context of an effective SHEQ&S program based on metrics which affect process safety performance. If any of these mapping tools already exist in some form in a company, the program integration team should use them first as the foundation. Once the program integration team has created its visual diagrams, it is recommended that others not directly involved in developing the mapping tool review and verify the team's diagrams. A successful mapping diagram should be understood by someone unfamiliar with the process being depicted.

4.9.1 Process maps or relationship maps

Process or relationship maps show an overall view, based on the departments or groups in an organization. Although they are often created to show how the different groups interact with suppliers to the organization (the input) and how they interact with the organization's customers (the output), for the purposes of this guideline, the "input" will be the metrics which affect process safety performance and the "output" will be the results from the metrics analysis that shows a performance improvement, as applicable, to each SHEQ&S group.

A process map is shown in Figure 4-22. When this process map is overlaid onto an organizational chart, as is shown in Figure 4-23, the information flow for the process unit metrics begins at the process unit level, is aggregated for each facility, and then each facility's metrics are aggregated for the corporate level. By design, the SHEQ&S program process map should overlay the company's management structure.

An example of a similar metric between the process safety and environmental groups could be as follows:

Process safety group metric: "the number of loss of containment events" (focusing on hazardous materials)

Environmental group metric: "the number of spills" (ranging from a minor leak to a significant release)

In this case, the SHEQ&S program would focus on the hazardous materials which have a process safety-related consequence, such as flammable liquids with the potential to form an explosive vapor cloud, by effectively combining metrics into one process safety-specific metric.   Recognize that there may be other materials with environmental permitting risks, only, which must continue to be addressed by the Environmental group.

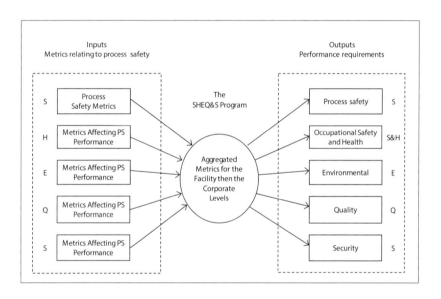

Figure 4-22. The metrics "process map" for the SHEQ&S program

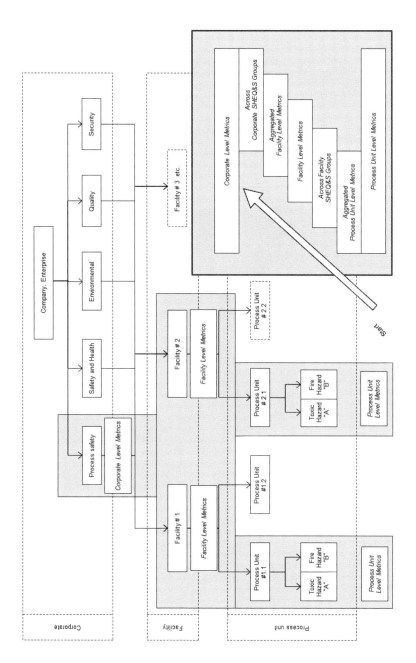

Figure 4-23. An example of a process map for metrics in an organization

4.9.2   Swim lane charts or cross-functional maps

Swim lane charts or cross-functional maps or show which group or department performs each step with the inputs and outputs for each step. These maps have more detail than a relationship map, but less than a process flow chart. The benefits for this overview level chart, located in between the less detailed process map and the more detailed process flow diagram, includes which group is responsible for what and how an organization ensures accountability for the measurement and tracking of the metrics. For example, those at the process unit level managing process hazards are responsible for ensuring proper operation and maintenance of the equipment, those at the facility level are responsible for ensuring proper resources to operate and maintain the equipment, and those at the corporate level are responsible for ensuring proper resources to run the facilities. An example cross-functional chart is shown Figure 4-24.

Inconsistencies between what is expected and what is actually done become evident in these charts, as the inconsistency shows up as a "mismatch" on the organization chart. For example, personnel at the corporate level have no role in developing, have no line management authority over, and have no oversight for the facility-level procedures. Although these charts will show duplications, nonproductive steps, and gaps in coverage in the management system, they may take a long to time develop and may be difficult to read. To help avoid confusion, complex charts need to be simplified when used to communicate the SHEQ&S system integration efforts, especially when working to secure support across the different SHEQ&S groups. The chart in Figure 4-24 could be used as a simplified chart before presenting the actual, more complicated reporting structure of the company.

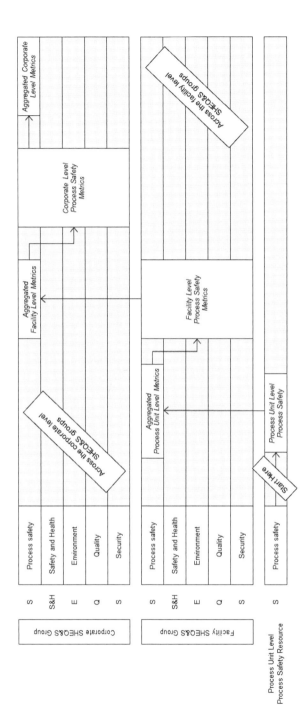

Figure 4-24. An example of a cross-functional chart for integrating metrics in an organization

4.9.3 Process flow charts or process flow diagrams

Process flow charts or process flow diagrams take a single step from a swim lane chart and expand it to show more detail. When focused on the management process, this level of detail can provide a clearer picture of the steps in the existing metric monitoring and reporting process. If a process flow diagram specifies the tasks, it helps identify problems such as bottlenecks, repeated steps, missing steps and so on. Although process flow diagrams help in quality-related continuous improvement efforts to reduce cycle times, avoid rework, eliminate inspection or quality control steps, and prevent errors, care must be exercised to ensure that the steps affecting how process safety risk is managed are not compromised when making changes. These process flow charts focus on the material and information flows throughout an organization, differing from the process map which is more focused on the personnel who have control over and directly influence the resourcing decisions.

With the goal of helping improve the work flow and reduce the work load on SHEQ&S resources, the chart from Figure 4-24 is rearranged in a process flow diagram and shown in Figure 4-25. Each facility will have identified its metrics using process unit-specific resources, with each level in the organization aggregating them through the facility and corporate SHEQ&S groups. T his combination procedure, the aggregation of the process unit level metrics for monitoring at the facility and corporate levels, is similar to the procedural "nesting" concept, where each step described at a higher level can be broken down into more detailed steps at the lower levels. The diagram shown in Figure 4-25 can be used as a framework for more detailed aggregation of the process safety area metrics in an organization. The process safety areas inherent for successfully managing a general process safety program were shown in Figure 4-17, with the specific CCPS RBPS approach delineated in Figure 4-18 and Figure 4-20, respectively.

From the quality management systems point of view, there are four continuous process improvement phases in these levels of analysis that evaluate the detailed process activities and flows by:

1. Analyzing each process step for:
   1.1. Bottlenecks
   1.2. Duplications
   1.3. Rework (fixing instead of preventing errors)
   1.4. Unnecessary steps
   1.5. Sources of delay
   1.6. Role / responsibility ambiguities
   1.7. Cycle time
2. Analyzing each decision for:
   2.1. Authority ambiguity
   2.2. Are the decisions needed at this point?

3.   Analyzing each rework loop to:
   3.1.   Prevent the rework step(s)
   3.2.   Eliminating steps
   3.3.   Do the rework in less time
4.   Focusing on the customer's point of view (a distinctive "quality system" aspect):
   4.1.   Value-added vs. non-value-added steps (with the ultimate "customer" being the company)

Although the benefits process flow charts conform to the ISO 9000 requirements, it may not be easy to identify some types of the system-related process breakdowns, such as identifying those responsible for the step or when steps inadvertently skip personnel who need to be a part of the decision making process.

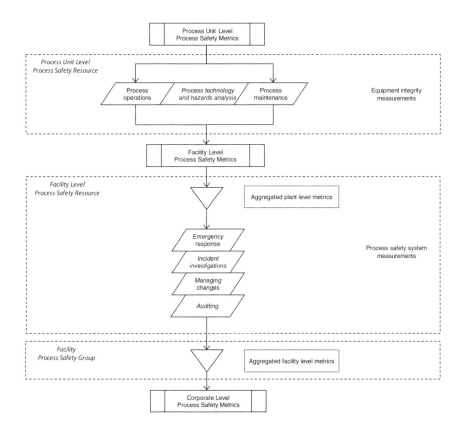

Figure 4-25. An example of a process flow chart for aggregating metrics from the process unit level

4.10   OTHER METRICS WORTH CONSIDERING

If the metrics selection team is having difficulty identifying particular metrics that affect process safety performance which can be integrated between the different SHEQ&S groups, there are many available resources that provide guidance on potential metrics [ACC 2013c, API 2010, CCPS 2010, CCPS 2011, HSE 2006 and OECD 2008].   Leading and lagging indicators could vary within the same company between facilities at different locations based on their system maturity, the length of time working on the systems, and their frequency of measuring and reporting.

5 IMPLEMENT THE SHEQ&S PROGRAM

A management framework for integration provides the skeleton on which the SHEQ&S program can be built and implemented. This framework helps define the overall structure of the integrated system, the way it is built, the sequence in which it is built, and which tools can be used to build it. Correctly designed, the framework helps ensure that the program integrating the systems matches, if not enhances, the current management systems. This chapter describes some approaches that will help with the implementation effort. Specifically, this chapter describes the implementing phase, the "install and test" phase, of the project, including approaches to help pilot the program for this integration effort, as is shown in Figure 5-1.

At this phase in the integration project, support for this integration has been obtained, the existing management systems have been identified, and the common, risk-ranked metrics which affect process safety performance have been identified using the tools provided in Chapter 4. It is worth reemphasizing that this is not a new initiative. The SHEQ&S program is taking advantage of the existing management systems and is providing the company with a more effective approach to managing its operational risks. This guideline has described a SHEQ&S program in terms of a "project" to help illustrate the phases essential for a successful program implementation. The resulting program is a process – it does not have a defined end point. Implementing the SHEQ&S program is just a part of the continuing journey to better manage and reduce process safety risks.

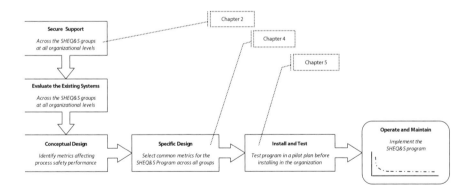

Figure 5-1. The install and testing phase when implementing the SHEQ&S program

This chapter first discusses how to apply the PDCA approach when implementing a SHEQ&S program, how to prioritize the integration efforts, how to develop integrated systems, and then how to build the concept of continuous improvement into the system's life cycle. Before implementing a full SHEQ&S program, an organization must develop a pilot plan to test the system on a limited scale to learn what does and doesn't work. Set in context of the sections in Chapter 4, the framework of this chapter is shown in Figure 5-2. The piloting section in this chapter provides ideas that should help ensure its success, taking advantage of both existing and informal management systems along the way.

At this point, the metrics selection team has selected the common metrics for the SHEQ&S program and determined how they are currently managed across the groups. Significant differences between the management systems, if they exist, need to be taken into account in the design of the program. Since there are many different organizational structures with existing and informal management systems, this guideline cannot address all the implementation permutations that could exist. Each organization needs to assess which of the different, existing systems being used to perform the work (its work processes) are the best, and select among or combine the best systems for the SHEQ&S program. The purpose of selecting metrics common to each SHEQ&S management systems is not to add more work but to enhance the current work processes. This chapter can provide readers with ideas on how best to apply the concepts during their implementation efforts.

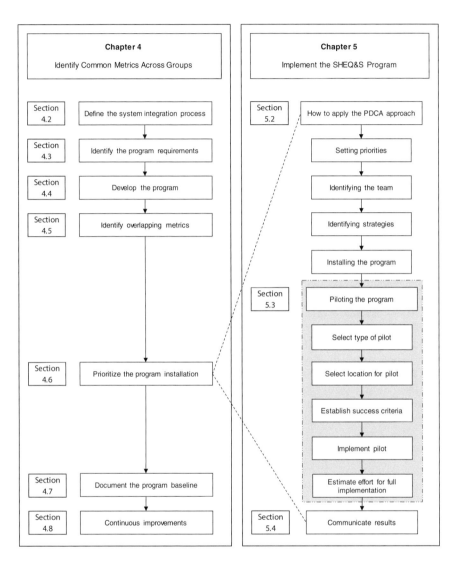

Figure 5-2. The framework for implementing the SHEQ&S program, including its piloting effort

5.1   THE NEED FOR PROPER IMPLEMENTATION

The need for an effective SHEQ&S program implementation hinges on two facts: 1) stakeholder's tolerance for catastrophic accidents will continue to decline, and 2) meeting or exceeding stakeholder's expectations for safe operations will continue to be crucial for long-term company success [CCPS 2007a]. T he demand for safety, health and welfare of people and the environment from external stakeholders will continue to grow, especially as the digitized information flows across the internet, providing live, incredible visual images of fires, billowing black smoke or oil flowing out of underwater wells [Vaughen 2012]. V isible upper management leadership and support, along with everyone in the organization executing correctly every time, are essential for continuous process safety performance improvement.

With respect to everyone in the organization executing correctly every time, there has been considerable attention in recent years focused on improving conduct of operations and operational discipline to help reduce process safety risks [CCPS 2011c, Klein 2011 and Vaughen 2011]. The goal for effective implementation is that the work processes resulting from the SHEQ&S program. The "way things are done here," are the work processes that help effectively manage a company's risks. Conduct of operations is defined as "the embodiment of an organization's values and principles in management systems that are developed, implemented, and maintained [CCPS 2011c]. Operational discipline is defined as the "performance of all tasks correctly each time" [CCPS 2011c], and as "the deeply rooted dedication and commitment by every member of an organization to carry out each of their tasks the right way, every time" [Klein 2005]. Organizations have shown improvements in performance across many different industries, such as the nuclear and aviation industries, when they address complexity and utilize the principles of high reliability when managing risk [Leveson 2011, Vaughen 2012, and HRO 2013]. W ithout proper implementation of an effective SHEQ&S program, an organization will not effectively improve its process safety performance.

5.2   HOW TO APPLY THE PLAN, DO, CHECK, ACT (PDCA) APPROACH

This section provides a r oadmap for the PDCA approach when planning and implementing the SHEQ&S program. The planning must address the need for the effort, the organization's readiness to implement the system and be clear on its scope and who will be affected. The organization's policy, supported by leadership at all levels, includes ensuring the safety, health and well-being of people, both inside and outside the facility fence line. The SHEQ&S program will constantly evolve, responding to both internal and external pressures over time with its focus on improving the organization's process safety performance.

Emphasizing the SHEQ&S program life cycle's evolution over time (its continuous improvement expectations), these four components fit into the PDCA approach as follows:

1) Planning *(described in Chapters 2, 3 and 4)*

The plan is designed to fulfill the organization's policy. Improvements in process safety performance, based on effective use of the organization's SHEQ&S resources, includes understanding and controlling the process hazards, their associated risks and impacts, both inside and outside the fence line. Planning is an *ongoing process* which can be impacted by numerous internal and external events and activities.

2) Doing *(described in this chapter)*

For effective implementation, operation and accountability of its SHEQ&S program, an organization develops its leadership capabilities, its management support systems and its resources needed to achieve its policy, objectives and targets. An organization focuses and aligns its people, systems, strategy, resources and structure in order to achieve these goals, emphasizing the principles of conduct of operations and operational discipline which apply on a day-to-day basis for everyone in the organization. Implementation is a dynamic *continual improvement* process, evolving as gaps and opportunities are identified, reviewed and used.

3) Checking *(described in Chapter 6)*

For effective integrated SHEQ&S performance evaluations, an organization measures, monitors and evaluates its metrics, comparing actual performance against the objectives and targets. Systems for *continuous improvement* must be in place as well for detecting gaps, responding to and implementing appropriate solutions, with actions being preventive, corrective, or a combination thereof.

4) Acting *(described in Chapter 7)*

At appropriate intervals, all levels of the organization's management conducts reviews of their metrics (or the metric's aggregates) and of their SHEQ&S management systems to ensure satisfactory operation and to promote *continual improvement*. The scope of this auditing approach needs to be broad enough to address the dimensions of its SHEQ&S activities and programs. Note (again): Caution must be taken when assuming that aggregating metrics provide a clear picture of what is going on. Important details may be overlooked, such that important process unit specific responses may be warranted and will not be acted upon since they became "rolled up."

Utilizing the PDCA approach introduced in Chapter 1, Figure 1-3, and combining the quality-based management components with the system's life cycle concept, helps describe the effective components of the SHEQ&S program, as is shown in Figure 5-3. The Plan, Do, Check and Act components, emphasizing the hazards and a risk-based approach, have been introduced in this guideline and are described in greater detail elsewhere [CCPS 2007a].

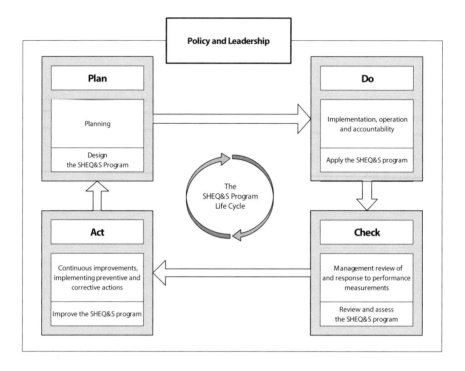

Figure 5-3. The four components in an effective SHEQ&S program

5.2.1   Setting the SHEQ&S program implementation priorities

Although there are no hard and fast rules for deciding the order in which to install programs and elements, there should be a plan to develop and install the management processes first. Without the management processes, the programs and elements will work in a vacuum, most likely being ignored by those who are supposed to use them (it is "business as usual"). Using the "priorities" noted for elements in a quality-based management system, the "categories" for selecting priorities for the SHEQ&S program installation process can be summarized in Table 5-1 [ISO 2013]. H owever, it is important to recognize that the actual priorities in any particular case vary depending on the facility's management structure and its culture. No matter how well planned and designed a management system might be, its implementation will fail if the culture is not ready for it.

To help with the integrated system prioritization, these categories are connected in Table 5-2 to the stages of the processing equipment's life cycle shown in Appendix E. This is consistent with the phases in common equipment maintenance, reliability and integrity efforts, with details of these "dependability" equipment-focused programs described elsewhere (IEC 2013). D ependability covers the availability performance of the process equipment. T he equipment performance factors considered in these efforts include: reliability (operations), maintainability (during its useful life), and maintenance support (maintain, remove). The process equipment that a SHEQ&S program focuses on is designed, fabricated, installed, operated and maintained to handle hazardous materials and energies. Incidents occur when the materials and energies escape from the process equipment and piping designed to manage them, resulting in a loss of containment.

Note that Table 5-2 also shows the process safety areas associated with these categories relative to the equipment life cycle. Although there are interactions between other process safety areas, the major systems identified among the categories are: conduct of operations, managing changes, equipment integrity, auditing, training and contractors. The change elements in Table 5-2 have been highlighted to reinforce the management of change element inherent in the evolution of any system: continuous improvements rely on change; managing process safety risks requires effective management of change systems to manage the risks associated with equipment-related changes. H ence, it is essential for improved process safety performance to ensure that a r obust "change management" system exists for the processes and the processing equipment, whether the SHEQ&S systems become integrated or not.

As an example for prioritizing and selecting among the SHEQ&S programs, a pilot could focus on the management of change process across all groups, addressing the integration exercise itself - continuous improvements inherent in the PDCA and systems life cycle approach. B y first integrating separate change control mechanisms into one robust, simpler change management process, the likelihood for successful integration of subsequent controls increases. This pilot test can provide immediate and visible results across the groups, including more rapid and less costly handling of changes incurred both internally and externally. Additional discussion on piloting the integrated system is discussed in Section 5.3 below.

Table 5-1. Possible SHEQ&S program installation priorities (shown in decreasing order of priority)

	Categories of requirements (i.e., through an ISO 9004 Management System)	Discussion on the category requirements
1	Management Responsibility Quality System Principles (System Structure)	These are fundamental to any quality-based management system and must be the first to be installed.
2	Design Control Purchasing Process Control Production Control	These help ensure that the equipment handling hazards materials and energies is correctly designed and operated. Other elements listed below help ensure compliance with this group's output.
3	Equipment tests and inspections Equipment used for the tests and inspections (see Category 8)	Safe operations depend on reliable process equipment <u>and</u> the reliability of the equipment <u>used to perform</u> the tests and inspection on the process equipment, as well.
4	Auditing	While inspections verify the processing equipment's conditions (the engineering design), auditing verifies the procedures and management systems (the administrative controls).
5	Nonconformity Corrective Action (refers to Categories 2, 3 and 4)	The tests, inspections and auditing systems must have effective systems designed to detect for gaps and correct for deficiencies (what is done when the results exceed acceptable tolerances).
6	Training	The development of training programs depends on the new or enhanced training requirements, and is continuous process, often requiring "refresher" training at specified frequencies.
7	Material control and traceability	If raw materials or material losses represent significant safety or environmental hazards and risks, this category will need a higher priority.
8	Quality records Use of statistical methods (closely tied to Categories 3 and 11)	These help document and establish performance measurements which are analyzed to identify address gaps and continuous improvement opportunities.
9	Economics	No quality system makes sense unless it is contributing to improvement in economic performance. However, until the bulk of the integrated system is in place it cannot be effective.
10	Contract review	When contract labor is routinely used, whether for equipment maintenance, equipment tests and inspections or capital projects, these reviews help ensure that contracts address SHEQ&S risks.
11	Quality documentation and records (see Category 8)	These must be developed in conjunction with the records that are used for validating the results of equipment inspections or the administrative systems audits.
12	Handling, storage, packaging, and delivery After-sales servicing Product safety and liability Purchaser supplied equipment	If applicable, these are important if the products fall under "cradle to grave" or "sustainability" programs where such responsibilities are expected by the manufacturer outside of its facility/plant fence line (e.g., Responsible Care®).

Table 5-2. A connection between the equipment life cycle and the process safety areas when considering the SHEQ&S program installation priorities

Categories of requirements (i.e., through an ISO 9004 Management System)	Equipment Life Cycle Stages							Process Safety System (see note)
	1 Design	2 Fabricate	3 Install	4 Operate	5 Maintain	6 Change	7 Remove	
1 Management Responsibility / Quality System Principles	✓	✓	✓	✓	✓	✓	✓	Conduct of Operations
2 Design Control / Purchasing Process Control / Production Control	✓	✓	✓	✓		✓	✓	Managing Changes
3 Equipment tests and inspections (See Category 8 below)	✓			✓	✓			Equipment Integrity
4 Auditing	✓	✓	✓	✓	✓	✓	✓	Auditing
5 Nonconformity Corrective Action (See Categories 2, 3 and 4)				✓	✓	✓		Auditing
6 Training				✓	✓	✓		Training
7 Material control and traceability						✓		Equipment Integrity
8 Quality records / Use of statistical methods	✓	✓	✓	✓	✓	✓		Auditing
9 Economics				✓	✓	✓		Conduct of Operations
10 Contract review	✓	✓	✓		✓	✓	✓	Contractors
11 Quality documentation and records (See Category 8)	✓	✓	✓	✓	✓	✓	✓	Equipment Integrity
12 Handling, storage, packaging, and delivery / After-sales servicing / Product safety and liability / Purchaser supplied equipment	✓			✓		✓		Conduct of Operations

Note: *The process safety areas include other elements (or pillars) that are essential for effective management of hazardous materials and their associated risks [CCPS 2007a, Sepeda 2010]. The systems identified above represent the "major" process safety element that applies. Process design and equipment design, in particular, will affect the layers of protection required for safe and reliable operation.*

In addition, setting priorities for developing and installing programs and elements should consider identified gaps in compliance or process safety-related risk management efforts, the likelihood of success (or the degree of difficulty), and the existing SHEQ&S management system strengths and weaknesses. Additional discussions for these considerations are as follows:

• *Addressing compliance or risk management gaps*

If regulatory compliance requirements are not being met or process safety risks are identified which are not effectively controlled, these issues must be addressed first, with the programs that correct the gaps given the highest priority.

• *Likelihood of success and degree of difficulty*

It is important to gain some early success that provides credibility to the integration project. These successes can be selected and planned, however "easy" targets may be dismissed as being unrepresentative. Balancing the selection between a mixture of easy programs and elements with some offering a greater challenge will build credibility. In addition, by working on some challenging systems early, problems and issues can be identified and overcome early in the project, harnessing the initial burst of attention and energy.

For example, integrating risk based hazards assessment programs might be the first to pilot. Although an effective risk based process hazards and risk assessment program may exist within the process safety group, the risk based approach may not be of comparable quality in the occupational safety or environmental programs. Developing an integrated program provides early benefits to these groups by using the existing process safety program as a basis.

• *Existing SHEQ&S management system strengths or weaknesses*

A strong program or element is generally well understood by everyone. This provides an interesting challenge—the difficulty in introducing change to something which is already working due to the reluctance to "fix something that isn't broken." However, even strong programs and elements may benefit with minor changes. Assessing the degree of change between the program's current state and its future state will provide a gauge on how much effort is needed to affect the change. The integrated system, by design, will help a weaker program bridge the gap effectively, providing more credibility to the program manager who will leverage other programs identified in the integrated system without having to invest significant resources from their group to affect the change.

For example, there may be different internal and external audits conducted by specialists within process safety, occupational safety and health (including industrial hygiene), environmental, quality and security groups that ask the same (if not identical) questions to the same people at a f acility. T he integrated audit program offers managers fewer audits, is less distracting to the staff, and will prevent duplication of effort to address deficiencies identified during the audit.

5.2.2 Identifying the program integration team members

Once the order for developing the SHEQ&S program management programs has been established, program integration team members must be identified and assigned work responsibilities. Although sharing resources in the SHEQ&S program, by design, will reduce the work load across each group in the long run, there needs to be a concerted effort initially from all groups to develop the integrated system. After identifying dedicated team members, the integrated SHEQ&S project needs to be formally chartered, resourced as needed, piloted, and then fully implemented. Initial responsibilities may be shared between the project team and managers at all levels during this phase, with special resourcing including people at process units that handle the hazardous material and energies. Smaller, independent development sub-teams may need to be chartered, as well, to focus on facility- or process unit-specific programs.

The members chosen for an effective program integration team should mirror the organization's structure. T he team should include, as appropriate, members from each of the process safety, occupational safety and health, environmental, quality or security groups who have one or more of these capabilities:

- Risk evaluation expertise and experience in their respective group
- Experience assessing these risks in their group's relevant technical disciplines
- Experience in relevant management systems (including developing systems, especially quality-related management systems)
- Experience in relevant operations activities
- Experience in relevant process equipment inspection and maintenance activities

For effective meetings, especially since the members will be from different levels in the organization, the team should establish its "ground rules," identify members with meeting facilitating skills, agree upon the division of labor and select a representative from the team who acts as the team's leader. T he team's charter should establish clear instructions on the scope of the work and should define the schedule for piloting and fully implementing the SHEQ&S program. The team leader will be responsible for compiling the discussions and for periodically reporting on the team's progress to upper management.

5.2.3   Developing installation strategies for the integrated system

Once the SHEQ&S program has been developed and reviewed with each group, the system is ready to be piloted successfully and then installed across the organization. After the successful pilot(approaches described below in Section 5.3), additional people may need to be enlisted during the installation phase, since installation is best managed by those who obtain the metrics and use their results to make decisions within the integrated system. Depending on the type of the change, interim processes may be needed to bridge gaps during the changeover from the existing to integrated systems. Although the interim processes may involve a significant level of effort and may take time, investing in the transition process is essential for success of the final system

A successful SHEQ&S program is reviewed, enhanced and approved by those who have to use it, with special training that explains the new system, how it benefits them, and how it will be used. Although there are many implementation strategies, three sample project installation strategies are shown in Table 5-3. Each example shows one possible implementation strategy. Variations on these will help meet local circumstances. Example 1 envisages shared responsibility for the project with local staff. Example 2 shows local staff taking the lead and Example 3 shows minimal involvement of local staff. Other combinations of responsibilities and the use of other resources are also possible. Constant in all the examples is the development of management processes ahead of programs and elements, and the provision of local training before installation starts.

5.2.4   Installing the SHEQ&S program

The management processes must be developed and installed before specific programs and elements can be installed. Although this may seem obvious, too many programs have failed when the "cart is placed in front of the horse." A management structure understood by all is essential for effectively managing the SHEQ&S program. Recall that guidance for establishing installation priorities was presented in Table 5-1, with the first class identified as essential – management responsibility and system principles (its structure). For any system or work process to be effective, clear responsibilities across groups must be established. This includes careful thought on who will collect the data, how the data will be collected and who will analyze it. Some useful management tools and methods are described in Chapter 4, such as the swim lane diagram used to "map" the work process, as was previously shown in Figure 4-23.

Table 5-3. Examples for strategies to implement the SHEQ&S program

Steps	Example 1	Example 2	Example 3
1	Integration Team develops Integrated Management Processes	Local staff trained on Integration approach	Integration Team develops Integrated Management Processes
2	Local staff training on new Management Processes	Integration Team and local staff develop Integrated Management Processes	Integrated Programs and Elements developed by Integration Team
3	Measurement of existing performance	Local staff develop integrated programs and elements	Interim arrangements developed
4	Integrated Programs and Elements developed by Integration Team and Local Staff	Measurement of existing performance	Local staff trained and proposed systems reviewed with them
5	Operator Training	Operator Training	Operator Training
6	Interim arrangements developed	Interim arrangements developed	Measurement of existing performance
7	Interim arrangements installed	Interim arrangements installed	Interim arrangements installed
8	New management processes installed	New management processes installed	New management processes installed
9	New programs and elements installed	New programs and elements installed	New programs and elements installed
10	Project review to assess pilot project	Project review to assess pilot project	Project review to assess pilot project

5.3   PILOTING THE SHEQ&S PROGRAM

At this point, the SHEQ&S program exists on paper and it is time to prove that the integration is practical, feasible, and beneficial. The effort required to implement new programs across whole organizations can be enormous. Wasted efforts, especially rework, cannot be afforded. A pilot implementation on a small scale provides an invaluable learning opportunity. During the pilot study, lessons are learned on a small scale such that changes to the system can be more easily made to address potential issues during full implementation. This section describes how to design a pilot study and how to learn as much as possible from the piloted exercise.

5.3.1   Selecting the type of pilot project for testing

The goal of testing the SHEQ&S program in a pilot effort is to prevent disruption to the existing management systems within and across the separate groups. Chapter 4 provided guidance on how to identify existing management systems used to measure the metrics which affect process safety performance across groups. These systems should provide sufficient information to help select a candidate type for the pilot study. Some criteria for selecting the types of pilot projects and for selecting where best to locate a pilot study are discussed in the following paragraphs.

5.3.2   The pilot should not be too simplistic or too difficult

By selecting a location with few SHEQ&S issues to manage overall or by narrowing the scope of the pilot to one specific program, the likelihood for the pilot's success will improve. Although the piloting effort may prove successful with a simple effort, the program integration team will have learned little to help identify and overcome potential problems that will arise with full implementation and most likely will not have improved their position to persuade any doubters to change their minds. A good example for a simplistic pilot is to select a warehouse that manages inventories of hazardous materials. Another good example for a simplistic pilot is selecting one process safety system only without addressing the complex system interactions, as is described in Section 4.6.1, Chapter 4. In either case, the learnings will probably contribute little to help with understanding potential issues that may arise during the full program implementation. Choosing a facility or department that requires a major piloting project with major efforts and changes may be too demanding, as well. An extremely difficult pilot may provide significant learnings at the expense of the program's credibility. A successful pilot will find a balance in between these extremes, not too simple and not too hard. If one exists, a facility or department that represents a smaller version of the full, completely integrated system would be an ideal choice.

5.3.3   The pilot should be able to measure improvement – some guidance

There is little point in conducting a pilot without being able to measure the improvements. Measures that can be used are discussed in this section, with clear goals for measuring success discussed in Section 0 below. Facilities or departments with good records for their SHEQ&S group's performance are good candidates, especially those with records that enable the efficiency of the existing management systems to be measured.

The data on improvement is critical for justifying the full project. If the results are inadequate, the pilot has not proven its benefits and there is a possibility that the full project implementation will not be authorized. However, it is unreasonable to expect that the full benefits of integration will be achieved during the life of the pilot project. Some benefits are gained only as the staff becomes familiar with the new integrated system. The program integration team may be able to use experience of other quality management projects to forecast the full impact of integration after some measureable improvements are determined.

Although the SHEQ&S program will have metrics for improving process safety selected for each SHEQ&S group, the CCPS has developed guidelines for the measurement of process safety performance that can *provide useful ideas* on which *key process safety metrics* to measure [CCPS 2010, CCPS 2011b]. The CCPS approach for developing performance measures combines quality management systems with robust statistical methods. This effort identified well-defined, measurable, and practical process safety indicators of process safety performance. In the spirit of continuous improvement, these metrics are periodically reviewed and updated to sustain their relevance and usefulness. Additional information on these metrics is contained in a separate guideline in the CCPS series [CCPS 2010].

When selecting appropriate metrics, the program integration team must consider the different types of metrics that can be used. The type of metrics can include in-process metrics, compliance metrics (both to internal standards or external regulations) and specific "loss of containment" metrics that measure effects to the health and safety of employees working in the process unit or to the environment. Since there are many possible metric combinations between the SHEQ&S groups, it is important to have the survey and analysis of the measurements currently monitored and tracked for the pilot baseline. The discussion and approach for identifying overlapping metrics which affect process safety performance was discussed in Chapter 4, Section 4.5.

Improvements in safety and environmental performance may come slowly. Although major accidents are rare, any reduction in the "black swan" incidents would be evident in the industry only after several years [Murphy 2011, Murphy 2012, Murphy 2014]. For this reason, there must be a combination of "leading" metrics in the pilot program to show that improvements are being made. Not only are leading indicators more proactive measurements, they help to identify weaknesses before an incident occurs. They can show improvements in a matter of months. Depending on the scope of the pilot plan, if it is scheduled for a year to "prove" itself, metrics that take years to show improvements will not suffice. A discussion on the recent advances in process safety metrics, including the distinction between leading and lagging metrics (the rare incidents) was provided in Chapter 1, Section 1.8. Hence, it is recommended that a balance between leading and lagging metrics be selected in the pilot, as well.

5.3.4   Selecting a location for the pilot

Selecting a facility open to the pilot will increase the likelihood for the pilot's success, as well as the likelihood for success in fully implementing the SHEQ&S program in the organization. This section describes some of the barriers that may need to be addressed before piloting the system at a facility and presents some facility- or department-specific features that can help improve the success of the piloting effort. For large organizations, designing a pilot that focuses on a small part of the organization makes sense. For small organizations, the scope may need to be bigger for effective measurement of the pilot's success, and it may involve the whole organization.

5.3.5   Addressing barriers for the pilot

The barriers or concerns to integration identified earlier in Chapter 2 influence the location for selecting a piloting effort. S ince most managers prefer to have someone else debug a new program before they have to use it, resistance is often high for pilot studies. The resolution of the barriers generally remains the same as for the full program. Depending on the size of the organization and the scope of the piloting effort, partial or total relief of the piloting costs and staffing at a facility may be absorbed through corporate.

If costs are any issue, it is important to emphasize that the process, personal and environmental safety management-related costs are costs that help reduce, if not prevent, the costs to operations associated with a process safety incident. The industry still experiences many preventable incidents due to inadequate hazardous materials management systems and programs. Compared to mature, effectively managed safety systems, relatively young safety management systems, whether recently introduced to the facility or created to manage new process technologies and processes, may be conservatively designed with too many resources allocated to manage the risks. These younger management systems have been over-designed in the name of "safety," making sure that there is a sufficient safety "buffer" for safe and reliable operations. However, the systems have not had the time to establish the resource optimization that minimizes the overall risk. The use of quality management approaches helps ensure that the new system is more efficient and costs decline over time.

Organizational issues may also be raised, as the integrated system may require changes in both the organizational structure and staffing. The selection of the pilot project must address some of these issues to be a cr edible test, yet too much organizational change makes it more difficult for the pilot study to be completed quickly and increases resistance. Therefore it may be useful to pick a location that has more flexibility in its organizational structure - either because it is smaller with individuals having diverse responsibilities or because the key individuals have very secure positions within the organization. An example success story is provided in Section 5.3.6 below, in which a slow and deliberate organizational change at a facility "transformed" itself into a proactive, safety culture, beginning at the leadership level. S ignificant performance improvements across several SHEQ&S groups at the same time occurred with the new culture.

5.3.6 Locating a facility for the pilot

In every organization, there are some departments that are more willing than others to try new ideas. S ome examples for identifying facilities which may increase likelihood of success for the integrated system piloting efforts are presented below.

Facilities with new leadership

Facilities or departments within a facility that have recently appointed new managers are likely candidates for the piloting effort. These managers may have inherited problems and old loyalties and often welcome a challenging project to bring their new team together. Note, however, that larger facilities with several operating groups must have a higher level sponsor with visible support to help protect the new manager with the new effort.

Facilities producing specialty products versus commodity products

Some facilities and departments work in fields where technology-related change is commonplace, and their staff thrives on this. For example, specialty chemicals operations introduce several new products each year and are familiar with, and relish, the challenge of change. Conversely, established facilities with static departments who have rarely undergone change may resist the idea of a pilot project. For example, a co mmodity chemicals operation focused on producing specified quantities of a few specific product grades may be reluctant to take any risks. ("Why fix what is not broken?") Wh ether it's a s pecialties- or commodities-based operation, successfully implementing new systems must address the work processes associated with the interactions between people (see the brief safety culture discussion below).

Facilities focused on quality management

Some facilities have implemented a quality management system efforts, such as those associated with the Six Sigma or Lean processes, that helps them maintain, if not grow, market share with satisfied customers. These facilities and its departments have a history of supporting and buying into quality concepts, since they naturally endorse continuous improvement. Their "new" customers will be internal, as the benefits of the SHEQ&S program will not be difficult to sell – the facility already knows that there is "always room for improvement." These facilities must provide open communications between everyone associated with the change, as technical and system advances risk the danger of becoming the "flavor of the month," losing their support where it is needed the most – at the floor.

Facilities with proactive, interdependent safety cultures

Locating facilities that have an open and proactive leadership and a strong, interdependent safety culture will help with the piloting effort [DuPont 2013]. Resistance to performance improvements across all groups is reduced when examples of success from other facilities are shared, such as the one presented in

Figure 5-5, when a significant culture change was implemented and sustained at a DuPont facility [Knowles 2002]. In this case, the DuPont Belle facility handling large volumes of toxic and flammable materials reduced its injury rates by 96%, reduced its environmental emissions by 88%, improved its productivity 45%, and its earnings were up 300% from 1987 through 1995. This effort was sustained for more than a decade when, unfortunately, management stopped talking to people after 2006, their words became inconsistent with their actions and retirements and staffing changes negatively affected those working on the floor. When combined with the complexity of managing process safety throughout the SHEQ&S systems, the gains which were built on the strong, trusting relationships across the organizational levels at Belle were lost. The loss in management support and interdependence led, in part, to a series of severe toxic release incidents in 2010, with one of the incidents resulting in a fatality due to phosgene exposure [US CSB 2011b]. The series of release incidents have subsequently resulted in costly fines by the EPA, as well, which were imposed in 2014.

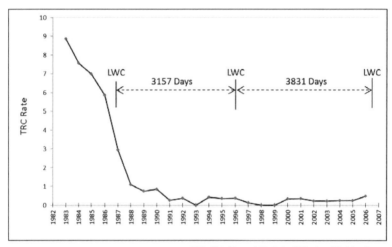

TRC = Total Recordable Rate; LWC = Lost Workday Case

Adapted with permission from Richard N. Knowles.

Figure 5-4. An example of how a culture change in the leadership improved metrics

5.3.7 Establishing the pilot's success and failure criteria

For establishing the success criteria of the piloting effort, a balanced selection between useful, measurable leading and lagging metrics (Sections 5.3.2 and 5.3.3) should be *combined* with a piloting scope that is not too easy or too challenging (Section 5.3.6). Developing and installing the integrated system, whether at the pilot phase or the full implementation phase, a balance of "early successes" and "challenges" is essential for demonstrating the benefits of the integration effort. If the stakeholders are involved in the selection of the success criteria, the pilot's success will have credibility for the piloting integration effort proceeding to the implementation phase of the SHEQ&S program.

By establishing the success criteria in the beginning, the piloting results will be more credible when presenting data that either supports success or presents its failure. The success provides the incentive for pursuing full implementation; the failure forces the program integration team to reassess the effort and learn from the experience, and if the approach is not salvageable, abandon the approach altogether (it's "back to the drawing board"). Remember that the purpose of a pilot is to "work out the bugs" of the SHEQ&S program on a small scale before the full SHEQ&S program is implemented across the entire organization. Part of the piloting effort is, by design, to quickly recognize unanticipated, potentially adverse consequences, and correct them when the consequences are small. In the process hazards analysis parlance, the pilot is designed to mitigate the consequences that could potentially occur if not addressed early in the implementation process.

The team should measure the success of the piloting effort itself, too. Progress must be periodically reviewed and compared to the piloting plan to ensure its success at the end.

Although one of the organization's goals is to ensure the safety, health and welfare of people from hazardous materials and energies, having proof of specific economic benefit to support the integration effort will make changes for full implementation easier to justify across the different groups in the organization, as well. Some of the management-level concerns noted earlier specifically address economic issues: "the implementation costs are too high," and "the cost reduction goal is not being achieved." Hence, the measures of success should include some economic benefit, as well. In essence, has the organizational work load decreased as a result of integrating the SHEQ&S system based on metrics which affect process safety performance?

5.3.8 Implementing the pilot

For successful implementation of the SHEQ&S program in the organization, the design of the pilot's installation phase should be similar to the design of the full project. The vision for the program was discussed in Chapter 2, Section 2.3, was presented in Figure 2-5, and was addressed with the stakeholder questions and answers provided Appendix C. The piloting lessons learned are more easily transferred and applied when fully implementing the SHEQ&S program. The piloting effort also provides the information and steps which can be used for training the rest of the organization upon full implementation. The piloting lessons address the bumps in the road on our journey for process safety performance improvements in a smaller but highly manageable way.

5.3.9　Estimating the level of effort for a fully implemented SHEQ&S program

Although the program integration team may be able to use prior experiences from other system implementation efforts to forecast the full impact and costs of integration, information may be available during the course of the piloting effort that can help with these estimates. Any issues identified during the pilot's implementing phase can be "forecast" for future efforts, depending on the size of the "bump in the road." The level of effort required varies dramatically depending on a number of factors, including:

- whether a particular aspect of integration is based on an existing system or if a completely new design is required;
- whether the existing system already aligns with the integrated management system's structure (more effort is required to bridge the gap); or
- whether the existing system that aligns with proposed management system structure contains all the essential elements for success (are additional features needed?).

Estimates for the level of effort combine the estimated number of people involved as well as the design time expected to implement the full system. Starting with established systems in one group that most closely match the final integrated system, the gaps that need to be addressed in the other groups will not be as difficult to identify and upgrade. However, if there is no model from which to build a foundation, the level of effort required for an organization may be considerable. It is also important to include discussion time needed with the stakeholders at all levels of the organization. Take into account the resources and time which will be needed to make sure that their needs are met, and include the time needed to develop training programs and then to provide each group with the training. By building on existing systems and leveraging these systems, the implementation costs will be reduced.

Although some responsibilities for the safety, occupational safety and health, environmental, quality and security groups may shift to different managers and staff, *this is part of the design of the SHEQ&S program.* The perceived "resentment" over losing or gaining responsibilities is a significant part of the organization's resource allocation optimization objective (refer to Chapter 2, Figure 2-4). When responsibilities do transfer, ensure that the handover is effectively performed, including having the newly responsible staff work alongside the existing staff under the old system with the existing staff retaining some of their responsibilities, having the existing staff work alongside the new staff under the new system with the new staff taking over responsibility, or a combination of both approaches. Alternatively, an interim management process may be developed to manage both the existing and the new staff during the transitioning time. The handover issues for large corporations will differ among facilities due to the different management systems and cultures that exist at each facility.

5.4 COMMUNICATION

Although communications within the piloting study are managed between limited groups of people, effectively communicating progress and issues during the course of the pilot in large organizations may be difficult. It is important to remind others not directly involved in the pilot's progress that the fully implemented system will arrive, with major bugs worked out, at some defined point in the future. These regular communications need to be tailored for the different audiences, whether they are at the corporate, facility or process unit level. Communications at the process unit level must cover everyone who will be affected on every shift. Emails alone do not suffice as effective communication. Regularly reminding others of the intent of the SHEQ&S program, especially how it will make their lives easier in the long run, keeps the attention focused on the efforts, reinforcing across the organization that this effort is important for the company's future success.

Communication subjects before piloting the integrated system could include the expected duration and schedule for the pilot, the type of pilot, and the location of the pilot. Subjects covered during the pilot include measurable progress against milestones, any issues being addressed and even unexpected benefits that occur. After the pilot is completed, depending on the piloting metrics selected at the beginning, the estimated costs and benefits for a full implementation, the lessons learned, and the changes that will help improve overall success should be communicated to enhance the full implementation effort.

6 MONITOR THE SHEQ&S PROGRAM PERFORMANCE

The first five chapters in this guideline have focused on the planning, implementing and applying phases of a SHEQ&S program using the PDCA life cycle approach. The SHEQ&S program will use metrics which affect process safety performance across the SHEQ&S groups, applying systems and programs to operate safe and reliable facilities managing the process hazards and risks.

This chapter discusses the next phase in the PDCA approach, the checking phase, which focuses on the management review of and responses to the SHEQ&S program performance measurements, as is shown in Figure 6-1. It is recognized that the maturity of an organization's management system and the metrics it uses will differ between organizations. Therefore, for organizations that may not have mature management systems in place, this chapter can be used as a starting point for helping monitor performance. For organizations that have well established management systems and metrics, this chapter may be used to provide ideas which may be used as enhancements to the current systems.

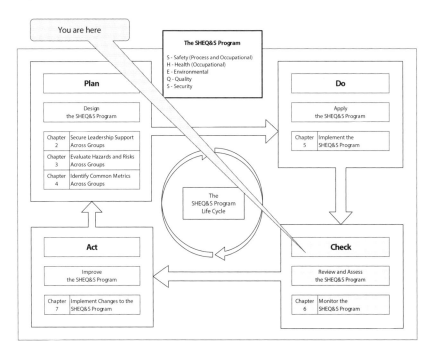

Figure 6-1. The review and assess phase in the "Plan, Do, Check, Act (PDCA)" approach

Since other CCPS guidelines already have been written to provide detailed "monitoring system performance" information, Table 6-1 is referenced in this chapter, as needed. The CCPS references in Table 6-1 are listed relative to the sections in this chapter. It is essential to note that the risk based metrics used to monitor the performance and efficiency must *effectively measure* the SHEQ&S program performance. Without properly chosen performance measurements, an organization will not be able to monitor and track performance improvements.

This chapter explores the monitoring and tracking framework essential for successfully improving process safety performance with the SHEQ&S program, provides a roadmap for these plans, and provides some ideas on how communications will vary depending on the stakeholder's level in the organization (i.e., the information they need and how their feedback should be used for improvements). It discusses a framework for monitoring the SHEQ&S program using leadership management reviews to respond to the deficiencies and gaps. Leadership must be engaged in this effort: auditing and verifying the SHEQ&S program then monitoring and tracking any corrective actions until the actions are effectively implemented and closed.

In addition, this chapter includes a short section that briefly describes some statistical methods and tools which should help with the data analysis and trending, and a section describing why it is important to "capture" success early: to establish credibility for the effectiveness of the SHEQ&S program in improving an organization's process safety performance. Examples of monitored, analyzed, and tracked metrics with their corresponding presentations used for communications at different levels in the organization are included in Chapter 8.

6.1   THE NEED FOR REVIEWING AND ASSESSING PROGRAM PERFORMANCE

The results from measuring, analyzing and tracking metrics lets an organization know if it is getting better in its process safety performance over both the short and long term. If the metrics do not measure the important process safety-related hazards and systems designed to manage them, then an organization does not know where its deficiencies are and what process safety risks actually exist. If it does not detect them, it cannot effectively respond to them.

As has been noted before, process safety metrics differ from the traditional occupational safety and health metrics and must be addressed specifically to improve process safety performance. Leadership needs to base its process safety risk reduction decisions off the combination of metrics data analyses, risk assessments, investigation root cause analyses, audit findings, etc. When a catastrophic process safety incident occurs, it may be a surprise to everyone in the organization, especially when there had been no history of it ever happening in their process unit before. Fortunately, recent advances in the types of measureable process safety metrics can proactively provide organizations with measures from which to select and review how they control and manage their process safety hazards and risks [CCPS 2010, HSE 2006, HSE 2011b].

Table 6-1. Other CCPS guideline references

6	Monitor SHEQ&S Performance	Reference	Specific Chapter
6.1	The Need for Reviewing and Assessing Program Performance		
Framework			
6.2	How to Reinforce the Integrated Framework	CCPS 2007 (RBPS)	Chapter 3, Process Safety Culture
6.3	How to Use Management Reviews to Respond to Gaps.	CCPS 2010 (Metrics)	Chapter 7, Drive Performance Improvements Section 7.5, Management Reviews
		CCPS 2011a (Auditing)	Chapter 23, Management Review and Continuous Improvement
6.4	How to Engage Leadership		
6.5	The Roadmap And Process Improvement Plan	CCPS 2007 (RBPS)	Chapter 22, Management Reviews and Continuous Improvements
		CCPS 2010 (Metrics)	Chapter 7, Drive Performance Improvements Section 7.5, Management Reviews
		CCPS 2011a (Auditing)	Chapter 23, Management Review and Continuous Improvement
6.6	Auditing and Verifying the Program	CCPS 2007 (RBPS)	Chapter 4, Compliance with Standards Chapter 21, Auditing
		CCPS 2011a (Auditing)	Chapter 2, Conducting PSM Audits Chapter 22, Auditing
6.7	Tracking Corrective Actions		
6.8	Some Tools (Statistical Methods)		
6.9	Capturing Early Successes		
6.10	Improving Performance in all SHEQ&S Management Systems	CCPS 2010 (Metrics)	Chapter 7, Drive Performance Improvements
		CCPS 2011a (Auditing)	Chapter 23, Management Review and Continuous Improvement
Communications			
6.11	How to Use the Information	CCPS 2007 (RBPS)	Chapter 5, Process Safety Competency Chapter 8, Process Knowledge Management Chapter 14, Training and Performance Assurance
		CCPS 2010 (Metrics)	Chapter 8, Improving Industry Performance Section 8.1, Benchmarking
		CCPS 2011a (Auditing)	Chapter 6, Process Safety Competency Chapter 9, Process Knowledge Management Chapter 15, Training and Performance Assurance
6.12	Obtaining Stakeholder Feedback	CCPS 2007 (RBPS)	Chapter 7, Stakeholder Outreach
		CCPS 2011a (Auditing)	Chapter 8, Stakeholder Outreach
6.13	Some Metric Analyses and Communication Examples	CCPS 2007 (RBPS)	Chapter 20, Measurement and Metrics
		CCPS 2010 (Metrics)	Chapter 6, Communicating Results Section 6.3, Different Audiences

6.2   HOW TO REINFORCE THE INTEGRATED FRAMEWORK

The management review program reinforces the SHEQ&S program framework using metrics which consider and complement the metrics identified for monitoring the performance of the SHEQ&S program. T hese metrics should evaluate the quality and dependability of the existing practices, identify repeat findings, and identify delays with scheduled activities closing the gaps. T hey should help monitor organizational performance by measuring the types and number of audit findings or measuring the number of incidents identified with SHEQ&S management system deficiencies.   The CCPS reference providing additional details on reinforcing the integrated framework with a r obust process safety culture is listed in Table 6-1.

Different metrics are used to describe past performance, help predict future performance, and encourage behavioral changes to improve an organization's conduct of operations and its operational discipline. T hese metrics identify the current performance and compare this measure to a s tandard tolerance which is expected for excellent performance. A ny deficiencies that are identified become the gaps that leadership can prioritize and address with corrective actions. These improvement opportunities reinforce the belief that process safety incidents are preventable.   Leadership must be committed to reinforcing its organization's process safety culture with visible support by allocating both people and capital resources, as needed, for implementing and closing the gaps with corrective actions.

6.3   HOW TO USE MANAGEMENT REVIEWS TO RESPOND TO GAPS

Process safety system performance management reviews complement formal process safety system audits by evaluating the management systems designed to control the process hazards and risks. A facility's internal routine management review evaluates whether its management systems are performing as intended and producing desired results efficiently [CCPS 2007a, HSE 1997, HSE 2013a, HSE 2013b]. A udits are systematic, independent reviews to verify conformance with prescribed requirements, providing a snapshot in time [CCPS 2011a], whereas management reviews include looking at metric analyses, audit results, incidents, process upsets, and employee surveys/comments, etc. B y design, a management review is more frequent and less formal than an audit, monitoring the "health" of the management systems. Management reviews help detect systems-related issues before they manifest themselves into problems.

Although there are management reviews at all levels in the organization, the corporate level reviews cannot address the day-to-day operations issues at its facilities, and must rely instead on the facility-level reviews to gauge the capability of its process units. SHEQ&S program management review efforts should focus on the facilities handling hazardous materials and energies. Management reviews should occur at specified frequencies, usually on the order of months, with well-defined procedures for documenting system deficiencies, generating recommendations, identifying deadlines, and assigning responsibility to individuals.

With the focus on safe and reliable operations, operational reliability plays an important part in providing information for these reviews. In addition, highly reliable organizations develop reviews in which everyone can raise concerns, especially on things they see as "unexpected," whether it involves controlling the process or if process equipment deviates from its design intent. These reviews include addressing issues raised by the users of the management systems, which need to evolve with the changing operational issues, as well. The CCPS references providing additional details on management reviews are listed in Table 6-1.

6.4 HOW TO ENGAGE LEADERSHIP

The senior leader should chair the management review meeting with his direct reports and other key staff members. Leadership at all levels in an organization becomes engaged in the management review process when they understand the answers to these questions:

1) What is the quality of our management system or program?
2) Is it providing the results expected?
3) Are we working on the right things?

By developing or selecting metrics for the management reviews, people at all levels in the organization can act on these reviews (within their jurisdictions) to implement changes and help improve performance. The output is an indicator of performance showing that improvements are being achieved, both for the short term and over the long term. If the quality is not being met, management has the answer for the first question. If the results are not as expected, then the second question is answered, and the answer to question number three is probably "no."

Engaging leaders at all levels during the new system training sessions provides an opportunity for going beyond the obvious benefits of improving knowledge and skills of the newly integrated system. These sessions provide opportunities to discuss the existing systems and help identify areas where opportunities for improvement exist. This training effort will provide future users with opportunities to help identify problems before the system is implemented.

6.5   THE ROADMAP AND PROCESS IMPROVEMENT PLAN

A roadmap for the SHEQ&S program review includes the following elements:

1) Defined roles and responsibilities
2) Established standards for performance, and
3) Validation steps for program effectiveness.

The roles and responsibilities provide clear lines of authority for making decisions including who is responsible and accountable for the execution of the management review system. The owners of the management systems being reviewed are responsible for gathering and supplying the performance-related information, with others directly involved in following the system's procedures supplying the implementation-related information. O bservations from this information should include the issues as well as suggestions for improvement.

The standards of performance for the management reviews should be the same as the standards for the management system being reviewed. T hese standards include the management review's scope and objectives, the frequencies and depth of the reviews, the requirements for resolving the findings, and how the reviews are to be documented.

The ways to improve effectiveness of the management review should focus on methods for improving the performance and the efficiency of the activities that support the management system. Some ideas for improving a management review program's performance and efficiency are provided in the literature (see Table 6-1 for the summary). S pecific examples include how to maintain a dependable practice, how to conduct the reviews, and how to monitor organizational performance.

It may be necessary to redesign or adjust measures from time to time to reflect permanent changes in operations and/or regulatory requirements. Targets or goals which are based on number of tasks completed or hours expended are most likely to need such adjustments. Differences between expectation and reality should be analyzed and the root causes of these identified. When appropriate, changes to the management systems should be developed and installed. The CCPS references providing additional details on continuous improvement plans and for driving performance improvements are listed in Table 6-1.

6.6   AUDITING AND VERIFYING THE PROGRAM

Auditing is designed to uncover successes and failures of the SHEQ&S program. Caution must be used when relying solely on audit results, for if the right data is not measured in the first place, then important gaps will not be detected. For example, if audits or metrics review only training records, they may only measure the quality of those who received training. There may be a significant number of people who have not received training, leaving an information gap that produces negative results.

As was noted earlier, there are two general "breakdown" categories discovered during audits: 1) the one-time problems that are a result of a single error in following the system, and 2) the systematic or repetitive issues that are the result of inherent weaknesses in the management system. Since audits are proactive reviews, the solution to one-time errors should be based on the potential consequences, such as an incident or non-compliance situation, or if there are other safeguards in place to catch the error before its potential consequence occurs. Systemic issues require reevaluating the current system, with solutions ranging from minor procedural changes, adding other safeguards, or redesigning a system.

Verifying the SHEQ&S program can be interpreted as determining whether or not the integrated system's performance is meeting expectations or established targets. If the expectations or targets are found to be overly ambitious, no corrective action may be required and the performance metric needs to be reevaluated. In addition, the validation measures will change over time when the assumptions used to establish them change.

Audits help validate the management review program effectiveness since the management review, if performing well, should identify issues before they are detected in an audit. However, audits should not be superficial. Audits can identify problems/issues with metrics, such as:

- Metric data not being collected properly (garbage in/garbage out)
- Excessive manpower required for collecting data
- Metric always indicates 100% or 0%
- Metrics data not collected for key risks or weak barriers

Incident investigations provide another means for validating the management reviews since gaps identified in the investigation can either be an isolated management system gap or a systemic gap that was not identified during the management review. The CCPS references providing additional details for auditing and verification are listed in Table 6-1.

6.7   TRACKING CORRECTIVE ACTIONS

There should be a process for tracking the progress for closing the corrective actions that address the nonconformities identified during the management reviews or audits of the SHEQ&S program. It is important to verify that the action is taken properly and addresses the original recommendation, adequately reducing risk.

High priority, high risk actions should be verified "complete" by an independent, qualified person.

Methods for tracking these actions include databases or spreadsheets that prioritize actions and report on their status. Whether electronic or not, an effective tracking system contains a description of the original finding (why it is considered a deficiency), a clear scope for resolving the gap, one person accountable for closing the gap (they may not actually perform the work but are responsible that resources are available to do so), and establishes a specific date when the finding must be closed. For a closed corrective action, the system must document specifically what was done and when it was closed. Depending on which level is needed to address the corrective action, the deficiencies and gaps identified from a management review or an audit should be categorized, entered into the list, and tracked through the facility or corporate corrective action tracking data base(s), as appropriate. Since not all metrics apply at all levels, an organization should distinguish where and how to best track these findings.

6.8 STATISTICAL METHODS AND TOOLS

Since there are different types of measuring and tracking indicators, there will be different types of statistical tools used to analyze and interpret the data. If inaccurate data is obtained, the decisions based on erroneous analyses may not be effective, with a possibility for the actions from the decision actually contributing to a *decrease* in the system's performance. Performance data analyses help identify areas needing improvement by spotting trends, helping identify root causes, and helping prioritize the actions which are required to address the root causes. Recognizing that there are other types of metric analyses and tools, these examples are provided as ideas to consider and use, if applicable.

Although some metrics may be easy to identify, establish and track, it is important to establish a balance between leading and lagging indicators. The differences among leading and lagging metrics are noted in Appendix B of this guideline and elsewhere [CCPS 2011b, Hopkins 2009]. The metrics are used to determine whether expectations were met and whether people are going to *and do respond* to correct the deficiency. It is preferred to use metrics that identify potential problems before a failure or incident occurs. Effective implementation of corrective actions and decisions based on the metric analyses will improve process safety performance, noting that the proactive actions from monitoring and responding to the results from the leading indicators helps prevent incidents which result in the lagging indicators.

There are other dimensions to these metrics, as well [CCPS 2011b]. The continuum of lagging-to-leading indicators may be described based on what is actually measured and who will use the information from which they base their decisions. The "what" may be an activity or an outcome metric, with the activity metrics used to track leading metrics (whether an action occurs or does not occur) and the outcome metrics (whether the action produced the expected quality or performance). The "who" is the intended audience for the metric's data analysis. The audience may be internal, with the results used to make decisions that help an organization manage its efforts and activities. E xternal audiences are used to publically demonstrate the organization's performance. S ome of the external metrics may be required by certification groups or regulatory agencies.

This section describes some of the forms that metrics can take, which affect the statistical tools used to analyze the data, providing trustworthy results for decision-makers. Although details for statistical variation based on the Six Sigma approach to improve quality are beyond the scope of this guideline, this brief discussion is included to help point the reader to additional resources, as needed.

6.8.1 Forms of metrics

The metrics may be expressed in absolute forms, as ratios or as indices. I n addition, the choice of the form depends on the metric's intended audience, whether it for information only or for someone to respond. An absolute measure is a simple count of specific activities or events over some period of time. They do not measure the quality of the activity or event and often are not useful when comparing across the organization. Ratios are normalized metrics that provide a better context for comparing results across different parts of the organization. Indices are numbers expressing a relationship to a scale or a number on a scale, expressing the value or level of the index relative to other numbers on the same scale. Both ratios and indices are useful for performance "benchmarking," helping decision-makers understand if they have an issue that must be addressed or not.

Good metrics allow for accurate and detailed comparisons, lead to correct conclusions, are well understood, and have a quantitative basis (i.e., can be statistically analyzed). They must be reliable, repeatable, consistent, independent, and relevant to the process or activity being measured. There should be sufficient data for the analysis to be meaningful and timely, providing information when needed. If applicable, they should be comparable with other similar metrics, should be appropriate for company and regulatory compliance, and should be appropriate for the audience. In all cases, they should be easy to use and periodically audited to ensure that they are meeting and continue to meet all of needs of their audience (i.e., the stakeholders and decision-makers).

6.8.2 Statistical tools

Statistical tools provide decision makers with the significance of the metric data analyses, helping answer the question: is the information significantly different than what is expected, and if so, by how much? T he larger the statistical difference, the more significant the gap, and the higher priority the finding is. Simply counting and reporting events over time using absolute metrics does not tell the whole story. Hence, absolute data is usually graphed over time to help those using the data understand if things are getting better or if they are not. There are many ways to "normalize" metrics, by including data such as:

- The number of personnel hours worked during a defined period (# of "exposure hours" / unit time).

- The total production volume produced during a defined period (# lbs/hr, kg/hr, bbl/day, gallons/day, liters/day, etc.).

- The number of production lots produced during a defined period.

- The number of scheduled equipment inspections performed on time.

Using only normalized data to evaluate process safety performance should be approached cautiously, as these "rates" may draw attention away from the details of the process safety metric and what it is actually measuring.

6.9   CAPTURING EARLY SUCCESS

Since the SHEQ&S program performance improvements may take a long time to become evident, early "successes" must be captured to sustain support from upper management.   Choosing baselines from the existing systems and including information from the management reviews and audit reports of the different SHEQ&S systems before implementation of the integrated system will provide the baseline for comparing and proving improvements.

The indicators used depend on the specific objectives identified for the SHEQ&S program. The "check" step should confirm that the expected benefits of an integrated system are being achieved. For example, the loss of containment-related lagging indicators can reflect the overall effectiveness of the integrated management system, showing a decrease over time. The leading indicators for potential loss of containment events, on the other hand, could detect potential breakdowns which could lead to loss of containment incidents or other unwanted events. Efficiency and cost measures can track the management system's performance. Some of these measures are suitable for routine weekly, monthly or quarterly reporting, whereas others are more difficult to quantify and might be developed only annually.

Since large process safety incidents incur significant costs, it is important to keep lagging indicators within the scope of the analyses. Early failures of the integrated system, such as when an unusual increase in significant incidents occurs, could indicate that the integrated system may be missing crucial steps and that the system design needs to be re-evaluated. Near misses are a leading indicator for accidents and incidents and should not be neglected.

6.10 IMPROVING PERFORMANCE IN ALL SHEQ&S MANAGEMENT SYSTEMS

The major objective of making, analyzing and responding to the process safety measurements is to help improve process safety performance. Weaknesses in the process safety systems can be identified and corrective actions taken. Process safety performance will improve with effective implementation of the corrective actions. The purpose of the SHEQ&S program is to improve performance across all of the SHEQ&S management systems by focusing on metrics which affect process safety performance, distributing the benefits across all groups and the organization.

The integration of the SHEQ&S management systems into a SHEQ&S program should produce performance improvements in each of the groups. Since there are always opportunities for additional, continuous improvements beyond the integration effort, part of the assessment should ensure that the integration does not inadvertently cause a group's performance to deteriorate. This could occur if there are organizational staff reductions beyond the staff responsibility reductions (the latter benefit may *due to the* integration effort). If the results from the monitoring, data acquisition and analyses identify significant gaps in the process safety systems, or as new external demands occur, such as new regulations, changes required to address the gaps may overload the existing staff managing the SHEQ&S program. The CCPS references providing additional details on improving performance are listed in Table 6-1.

6.11 HOW AND WHEN TO COMMUNICATE THE INFORMATION

The information and results from the process safety measure analyses are the "outputs" from the SHEQ&S program. The goal of effective communications is to harness the power of these results to make good risk based decisions for each SHEQ&S group at all levels in the organization. Sections 6.11 through 6.13 help address these questions:

1) How does an organization best ensure that it understands and uses this information as an indicator of its SHEQ&S performance? (Section 6.11)

2) How does an organization effectively communicate this information to all stakeholders, reporting the data in such a way that its decision-makers can trust the data from which they base their decisions? (Section 6.12)

3) How does an organization know it is improving, both in the short term and over the long term? (Section 6.13)

Before answering Questions 2 and 3, an organization must answer Question 1, exhibiting a certain "Process safety competency" based on the selected risk based process safety measures used in the SHEQ&S program. This is particularly important for answering Question 2 above, as making decisions without understanding what impact may occur may put the organization at greater risk. An organization with sufficient process safety competency hinges on where the organization is today with its combination of these three interrelated actions:

1) continuously improving knowledge and competency,

2) ensuring appropriate information is available to people who need it – at all levels, and

3) consistently applying what has already been learned.

The CCPS references providing additional details on how and when to communicate are listed in Table 6-1. For additional discussion on the metrics that can be used to help identify weaknesses in process safety knowledge and its use (or lack thereof) across the organization, refer to the discussion and tools using the references shown in Table 6-1, as well.

Management reviews should be utilized to communicate results. From the information presented in these reviews, management can make effective decisions across all of its SHEQ&S groups and improve its process safety performance. The following sections describe some of the implementation and communications issues, which, if not properly addressed, may adversely affect the effectiveness of the SHEQ&S program. This is especially important when the reported process safety information is being used to make decisions addressing an organization's risk.

6.11.1 Ensuring that the measures are taken and analyzed at proper frequencies

Whether the report is monthly, quarterly or yearly, the frequency for acquiring the data and for reporting it must be appropriate. Some information, such as releases exceeding reportable quantities, or explosions that cause fatalities should be immediately shared across the organization. However, this data is not statistically reliable when reported over short time scales (they are low frequency, high consequence events – the "black swans" – and the measure is a lagging indicator).

6.11.2 Managing measures that are not easily obtained in the beginning

Some measures may be ranked as "high risk measures" that are not currently being measured and, hence, may not be obtained easily. An example could be using "in service equipment failure rate" information while focusing on an "in process" equipment reliability measure. Although equipment reliability data is important, especially on equipment used to control hazardous materials and energies, if there is no current maintenance monitoring and tracking system in place that captures the data, much effort may be needed to acquire it. Noting that the failure rate data is important, the program integration team should recognize that this measure is a "lagging" indicator, as well. There may be other equipment-related measures that are being taken, and these may be used as "leading" indicators to monitor and analyze, helping the process unit proactively address issues before the equipment catastrophically fails. There may be systems currently measuring these leading metrics, such as DCS/SCADA systems [Broadribb 2009].

6.11.3 Managing measures that have long gestation times

Some of the measures may need many months, if not years, before enough data can be obtained for analysis. However, once the integrated system is established, information on the "high risk measures" that require longer periods for sufficient information and trending analyses should be introduced. If longer frequency/event type of data tracking is required on a shorter time frame, then statistical tools can be used which can "roll" the monthly data reporting requirement using data over 12- or 24-month periods.

6.11.4 Managing existing measures that are no longer deemed useful

If the existing measures are deemed less effective for helping improve process safety performance than previously believed, it is recommended that there be a few months' transition time, with the "old" and the "new" metrics compiled and communicated to the organization prior to removing the old ones from the reporting system.

6.11.5 Managing efficiency measures

Some measures will be chosen because improvements in these areas were part of the justification for the SHEQ&S program project. Calculating these measures generally may require specific data collection exercises and data analyses. If there is a relatively high cost associated with acquiring and preparing these measures, they should be selected and used prudently. In addition, the selected efficiency measures must have a baseline from which to be compared. In other words, what was the "efficiency" of the group's management system before the integrated effort versus its efficiency after its implementation.

For many of the efficiency measures, the relevant data may be collected during audits or by records analyses. It is likely that a company has an internal reporting cycle which helps coordinate these measurement efforts. Since it may not be possible to change the routine of the corporate auditing schedule just to meet the program integration team's needs, consider the auditing schedule as well when selecting the efficiency measures.

6.11.6   Managing over-enthusiastic data acquisition requests

Although it can be fascinating to collect as much data on process safety performance as possible, there is a diminishing rate of return if too many measures are selected for the reports. Too much time spent on collecting metrics can mean less time making sure that hazards/risks are managed and the process is running correctly. The risk based approach for identifying metrics which affect process safety performance is designed to help prioritize which measures to use and help gauge which to report. Certainly, it may appear that "more data" helps when making decisions, however, exercise caution with what is communicated throughout the organization. Publishing too much information may confuse the audience more than help inform them, and hence, they cannot make effective decisions with the information [Klein 2009].

6.12   OBTAINING STAKEHOLDER FEEDBACK

It is important to understand how the information is being used by the stakeholders and determining how well the SHEQ&S program is meeting the stakeholder's needs. It is essential that the data being collected is being used; there are too many stories from companies with people collecting data that was never used because "corporate required it." Hence, the better the aggregated metrics data is, the more likely it can be used and acted upon by upper management, whether it is at the facility or the corporate level.

Recall question two in Section 6.4: Is it what they expected and do they trust the information for effective responses? Information gathering, analysis and reporting problems include the following:

- Reporting information too quickly, preventing adequate time for data analysis, trending and responses

- Not reporting information quickly enough, preventing timely data analysis, trending and responses

- Reporting too much or insufficient information, such that the real message is lost in the noise or missed altogether

The stakeholders for the SHEQ&S program are those who are affected by and benefit from the integrated system. Their needs and concerns must be identified and addressed for the system to effectively reduce the work load across the organization and help improve process safety performance.

For the SHEQ&S program, employees at all levels in the organization supply and receive the information, making decisions based on the results of these metrics, whether they are "as measured" or combined data. The first measures are obtained at the process unit level, with aggregated measures supplied from the facilities to the corporate groups and the aggregated metrics supplied to external stakeholders, such as the owners, neighbors, local politicians and community leaders and regulators. The customers are those using the metric results to make decisions.

Stakeholder feedback answers also help answer the third question in Section 6.4: Are we measuring the right things and is it helpful? I t is important to recognize that information can be requested by the owner, people outside of the SHEQ&S groups, and external organizations, as well.

The two most common approaches to seeking feedback are written surveys and group meetings. Both cases may need a surveying specialist for designing the questions and conducting the survey. D on't rely on informal feedback. By deliberately designing and conducting well-structured and professionally designed stakeholder feedback surveys, the responses from these surveys will be useful for analysis.

In addition, some feedback information can be obtained at the time the data is gathered using a question in this vein: "Do you know why we are gathering this information and how this information is being used?" The answers to this question may be surprising and could indicate that the initial communications for the reasons why the data is needed were not effective or that those acquiring or using the information do not value the effort, possibly seeing little benefit for themselves or their group. CCPS has references providing additional details on stakeholder outreach (see Table 6-1).

6.13   METRIC COMMUNICATION EXAMPLES

Example process safety metric presentations designed to share the results from analyses at different levels in the organization are available on the CCPS website (refer to Chapter 8). Recognizing that there are other types of metric analyses and methods for communications, these examples are provided as ideas to consider, edit and use, as applicable. Quality analyses from measurement of the correct metrics will provide decision makers with trustworthy information.

For organizations with well-established process safety performance measures already identified, these approaches can provide ideas to enhance the existing processes, using the recent advances in the types of and analysis methods used for process safety indicators. F or organizations struggling with the best way to convey the analysis message to different audiences, these examples will provide ideas on approaches that have been effective in other organizations.

The CCPS references providing additional details on metric analyses and communication examples are listed in Table 6-1. These references include some example presentations, which are provided for communicating the metrics data to other groups, having been designed to "speak their language." When designing the communications for each level, there will be more process unit measures recorded and analyzed at the process unit level than reported to the corporate level. This is by design of the integrated management system, where process unit-level measures are combined for reports at the facility-level, and the facility-level measures are combined (aggregated) into reports submitted to the corporate level.

7 IMPLEMENT CHANGES TO THE SHEQ&S PROGRAM

The first five chapters in this guideline focused on the planning and applying phases in the SHEQ&S program's life cycle. Chapter 6 addressed the checking phase, where systems are in place to monitor and assess performance of the SHEQ&S program. This chapter completes the "Plan, Do, Check, Act" (PDCA) approach when integrating the SHEQ&S system: acting upon and implementing changes. T his action phase, or the "Act" phase, is shown in Figure 7-1. Continuous improvement is inherent in any quality management system, as there is always room for combining the learnings from the past with those from the present, then implementing changes tomorrow for better performance results in the future.

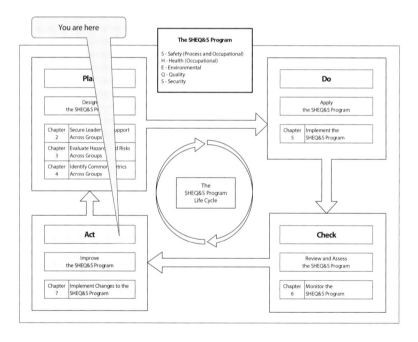

Figure 7-1. The continuous improvement phase in the "Plan, Do, Check, Act" (PDCA) approach

7.1   THE NEED FOR CONTINUOUS IMPROVEMENT

Continuous improvement is essential for the evolution and growth of the SHEQ&S program. Safe, swift and efficient continuous improvements can be made with an effective change management process. T hese continuous improvement efforts include improving the system's current management practices, addressing new issues with new responsibilities or practices, and reducing responsibilities by leveraging or removing a particular task altogether. A change management process reviews proposed modifications to understand the basis and objectives for the change, helping prevent new hazards and risks from being introduced or helping prevent changes that may compromise the existing management processes. As was noted in Chapter 6, continuous monitoring and assessments will identify gaps between the current and expected system performance. These gaps or "non-conformances" are addressed with corrective actions which make changes to the existing system. I n addition, the ongoing management reviews and audits help ensure that the system performance does not degrade over time.

7.2   ENSURING MANAGEMENT RESPONSIBILITY

Management is ultimately responsible for defining the objectives of any activity. Given that the "safety, health and welfare of people and the environment" is a core value in an organization, additional objectives may be formally established, such as "being recognized as the leader in safety, health and environmental performance in our industry" or "to be in compliance with all applicable regulations and industry standards." Part of management responsibility, therefore, is initiating any changes needed to keep the performance objectives on target.

Management is also responsible for ensuring that every stakeholder's (its "customers") needs are being met. The management at the process unit level will have to address deficiencies identified through the process safety metric data analyses. The management at the facility level will address process safety system deficiencies based on the aggregated metrics and the facility system metrics. Management at the corporate level will need to address the aggregated facility's process safety system metrics as well as its corporate process safety system metrics.

7.3   ADDRESSING NON-CONFORMITIES

There should be a process for addressing and correcting identified nonconformities and systemic gaps. This section discusses how addressing non-conformities is one of the aspects of a quality management system, the different types of non-conformities, the drivers of change that create corrective actions (including reviews that identify non-conformities), and guidance for creating a program that effectively manages changes required to address the non-conformances. The drivers and corrective actions associated with the process safety system "management of change" element is described briefly in Sections 7.3.3 and 7.3.4 below, with detailed references available in the literature [CCPS 2007b, CCPS 2008, CCPS 2010, CCPS 2013].

7.3.1   Non-conformance Evaluations are an Aspect of a Quality Management System

Addressing nonconformities is one of the key continuous improvement aspects of quality management system. The aspects, in the context of an effectively designed and implemented SHEQ&S program, include:

Management responsibility - Requires that each SHEQ&S group's management system is overseen by a manager and that each process safety system, both at the corporate and at the facility level, is overseen by someone. T his individual takes responsibility for initiating and following through to completion any improvement opportunities within their system (i.e., those generated by responses to non-conformities).

Personnel (training) - Ensures that everyone knows how the systems should work, that they know how to identify when it doesn't work as expected, and provides them with the tools or systems to identify and correct any underlying problems. Process safety investigations focus on identifying and understanding the "root cause" of the failure to meet the expectation or the deviation from the expected result (i.e., the non-conformance).

Product verification - Requires inspection and testing programs to confirm or verify that the "product" meets its expectations, specifications or targets. There are two "products" within the integration effort: 1) the implementation team's efforts to integrate the independent SHEQ&S management systems into one system; and 2) the expected performance of the metrics and their associated systems that are in place to manage and control the hazardous materials and energies. Non-conformities are identified when the verification fails.

Auditing - The quality system ensures periodic reviews of the system performance and that deficiencies are identified, and that corrective actions to address the deficiencies are created, approved, and tracked until closure. Management reviews are more frequent than audits. B oth the audits and reviews can identify non-conformances.

Use of statistical methods - Requires that the measured system performance be analyzed to identify strengths and the weaknesses. T he strengths of the stronger systems can be used as models for improvement; the deficiencies of the weaker systems will require actions for improvement. Statistical data analysis provides trustworthy non-conformance information analyses from which people can make effective decisions.

Nonconformance evaluations - Ensures that the root causes of any non-conformances (i.e., the deviations or failures) are identified and corrected, and that corrective actions are created, approved, documented, and tracked until closure. T he tests and inspections have defined specifications and tolerances with which to verify that the "product" is meeting expectations.

In addition to the quality management system aspects noted above, the SHEQ&S program must address compliance issues to demonstrate conformance with regulations and standards. T he periodic management reviews and audits described in Chapter 6 help identify such noncompliance issues, with corrective actions prioritized for prompt resolution. With the SHEQ&S program, these new external requirements can be quickly assessed across all of the groups to leverage potential changes to the other management systems.

7.3.2   The Different Types of Non-conformities

The different types of non-conformities or deficiencies are failure to meet a specification or target within a certain tolerance, failure to conform to compliance standards or industry guidelines, or a breakdown in the work flow within a management system. The SHEQ&S program will be affected by these gaps identified at all levels in the organization. Depending on the organizational level where the non-conformance is identified, different people will need to respond to the non-conformance. T hose at the process unit level address deficiencies identified through the process safety metric data analyses. Those at the facility level address deficiencies in process safety systems based on both the aggregated metrics and the facility system metrics. Those at the corporate level will need to address deficiencies in both the aggregated facility's process safety system metrics as well as its corporate process safety system metrics.

7.3.3   The Role of Non-conformances in Driving Changes

Non-conformances play a role as drivers for changes to processes handling hazardous materials and energies as well as the systems designed to manage hazards and risks. O ther drivers for both hazardous and non-hazardous process changes include:

- New process technologies or operating methods
- New technologies for existing process equipment
- New staffing or organizational changes (includes personnel substitutions and adding to or removing positions)

- Corrective actions from deficiencies identified in management reviews and audits, with priority given to regulatory compliance-related and safety performance deficiencies
- Corrective actions from incident root cause investigations, both from internally- or externally-shared incidents
- Corrective actions from updated hazard and risk assessments
- Corrective actions from equipment tests and inspections that fail specific tolerances or criteria
- New regulations or industry standards
- Insurance premium and cost increases
- Internal pressure to modify throughputs, improve efficiencies or improve product quality
- Internal pressure to reduce organizational costs

These changes range from minor changes in chemicals, technology, equipment, or procedures to large facility expansions or new facilities. The minor changes may be designated as temporary or permanent. S taffing changes may significantly impact process safety performance if they result in insufficient staff or staff with insufficient skills or training. Insufficient staff cannot be expected to operate, maintain, or support the process safety systems at full capacity.

Since each of these drivers for change may have some effect on each group in the SHEQ&S program, an integrated system will be able to make changes in one group and effectively manage the effects of the changes throughout all of the SHEQ&S groups. The SHEQ&S program change management process ensures that changes to one management system are effectively communicated and leveraged across other affected groups.

7.3.4 A Process for Managing Corrective Actions

A process is needed to manage the corrective actions generated by non-conformances. The major sources of continuous improvements on a sustained basis are the corrective actions from scheduled metric measurements, management reviews, and audits. C hanges to management systems occur when incident investigations identify a systemic root cause, as well. Wi thin the SHEQ&S program, the actual process for designing and implementing corrective actions can be shared among those who own the relevant programs, elements, and management processes that are undergoing the change.

The process safety system designed to manage changes, the management of change (MOC) element, provides a p rocess for effectively managing corrective actions. T he corrective actions (and the follow-up measures) are reviewed and authorized before being implemented. Process safety documentation is updated as a part of the implementation step for the change, with formal "pre startup" reviews performed before the change is implemented. The management of change process identifies who needs to be informed or trained on the change, defines what needs to be communicated (ranging from simple awareness to skills-based training), and helps establish the timing for the change. Specifically, the management of change element interfaces with other process safety systems. The primary interfaces to these systems [CCPS 2007a, CCPS 2010] are:

- Process knowledge management
- Hazard identification and risk analysis
- Operating procedures
- Safe work practices
- Asset integrity and reliability
- Training and performance assurance
- Operational readiness

These interfaces correspond to many of the quality system groupings previously described in Chapter 5. Hence, the management of change program addresses many of the quality system components included in the SHEQ&S program design.

Since the management of change program is an integral part of managing process safety and is used to identify, review, and approve all changes to process equipment, operating and maintenance procedures, raw materials or processing conditions, it will address changes that may affect the equipment life cycle, as well. The links for the management of change to the equipment's life cycle were described in Chapter 5, as well. B y designing and using the "management of change" process in the SHEQ&S program, changes affecting process hazards and risks can be effectively leveraged across SHEQ&S groups to help sustain, if not improve, an organization's process safety performance.

7.4 USING STATISTICAL METHODS

The statistical analysis of the SHEQ&S program process safety performance provides a powerful technique for continuous improvement efforts. When management monitors for and detects trends using statistical tools, trustworthy results can provide decision-makers with the information they need to make and implement effective changes that improve process safety performance.

8 EXAMPLES FROM INDUSTRY

Although the design of the chapters in this guideline focuses on selecting, monitoring, and responding to metrics which help improve process safety performance, over time it will become evident that everyone in the organization benefits when the company continues operations without accidents that harm people, harm the environment, or harm the continuity of operations. The selected risk based process safety performance indicators across the SHEQ&S groups have been properly prioritized, chosen, monitored, analyzed and are acted upon. Hence, an effectively implemented SHEQ&S program will help a company effectively manage its overall operational risk.

The examples provided in this chapter

1) Help answer the question: "Are we getting the results we want?"

2) Provide communication approaches for ensuring that the question posed above is effectively addressed across the company.

The results for each SHEQ&S group include improved procedures, streamlined audits, and compliance with local and governmental regulations. The associated corporate benefits include minimizing the load on the SHEQ&S staff at all levels in the organization, reducing operating and maintenance costs, improving operating and maintenance reliability, satisfying both internal and external customers, and improving overall company efficiency. These results and benefits to both the individual SHEQ&S groups and the overall company's operational risk reduction efforts can be depicted in Figure 8-1. Consequently, the management systems across all parts in the organization are more effective in helping staff at all levels in the company meet their objectives. The vision of the SHEQ&S program is realized once the program integrating the management systems has been implemented.

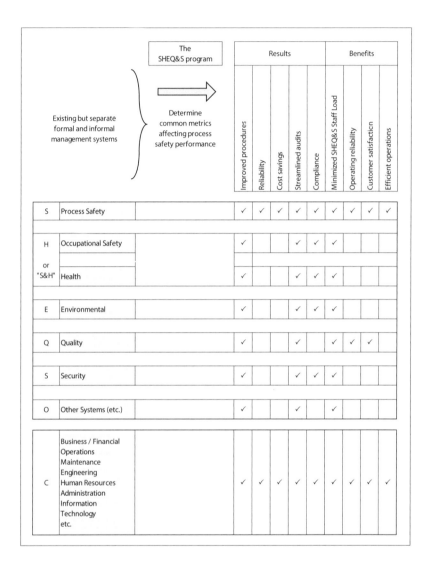

Figure 8-1. The results and benefits for everyone in the organization with an effectively implemented SHEQ&S program

As was discussed in Chapter 2, the general risk equation and general risk matrix were expanded to account for the overall operational risk to the company based on effective resource allocation across the SHEQ&S groups. When a company reduces the frequency and potential consequences of its process safety events, optimizes its resources, and the leadership drives the organization's process safety culture, instilling both the conduct of operations and operational discipline at all levels, the company will have reduced its overall operational risk [CCPS 2011]. This "optimum resource allocation" across the SHEQ&S groups is shown in Figure 8-2, where the people, the equipment, and the systems resources have been optimized. The company is not too liberal with its resources nor is it too conservative. The distribution is just right.

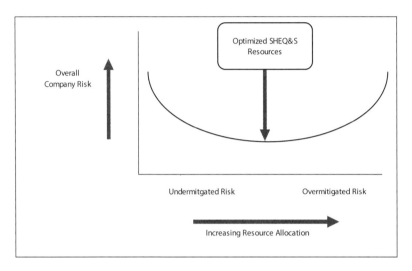

Figure 8-2. The overall company risk showing minimum risk when the SHEQ&S program is used that optimizes its SHEQ&S resources

8.1 CASE STUDIES

Case studies can be accessed on the CCPS website at:

http://www.aiche.org/ccps/publications/metrics-tools

8.2 EXAMPLES OF THE SHEQ&S PROGRAM

Examples of implemented SHEQ&S programs can be accessed on the CCPS website at:

http://www.aiche.org/ccps/publications/metrics-tools

8.3   EXAMPLES OF INTEGRATED AUDITING SYSTEMS

An example auditing tool for the SHEQ&S management system is provided in Appendix F: the "SHEQ&S Management System Mapping Survey."

> *The premise of the questions in Appendix F is to-*
>
> Successfully reduce the work demands on the different SHEQ&S groups by understanding and enhancing the existing management systems, not creating new work processes.

The questions posed in the SHEQ&S system mapping survey focus on the systems used to manage an organization's operational risk across the SHEQ&S groups. Since global organizations have facilities under different jurisdictions and regulations, its corporate standards and guidelines must be performance based, allowing each facility to develop their prescriptive, facility-specific standards and guidelines.

Other examples of integrated auditing systems can be accessed on the CCPS website at:

http://www.aiche.org/ccps/publications/metrics-tools

APPENDIX A: REFERENCE LISTS FOR GLOBAL PROCESS SAFETY LEGISLATION AND SHEQ&S ORGANIZATIONS

This appendix lists some of the global process safety legislation and SHEQ&S organizations at the time of publication. Please refer to:

Table A-1   U.S. regulations

Table A-2   International regulations

Table A-3   Voluntary industry standards

Table A-4   Consensus codes

Table A-5   Organizations Committing Efforts to Process Safety

Table A 1. U.S. Regulations

Process Safety	PSM - U.S. OSHA Process Safety Management Standard	Process Safety Management of Highly Hazardous Chemicals (29 CFR 1910.119),U.S. Occupational Safety and Health Administration, May 1992. www.osha.gov
	RMP - U.S. EPA Risk Management Program Regulation	Accidental Release Prevention Requirements: Risk Management Programs Under Clean Air Act Section 112(r)(7), 40 CFR Part 68, U.S. Environmental Protection Agency, June 20, 1996 Fed. Reg. Vol. 61[31667-31730]. www.epa.gov
	NEP - U.S. OSHA PSM Covered Chemical Facilities National Emphasis Program	PSM Covered Chemical Facilities National Emphasis Program, OSHA Notice, 09-06 (CPL 02), U.S. Occupational Safety and Health Administration, July 2009. www.osha.gov
	NEP - U.S. OSHA Petroleum Refinery Process Safety Management National Emphasis Program	Petroleum Refinery Process Safety Management National Emphasis Program, OSHA Notice, CPL 03-00-010, U.S. Occupational Safety and Health Administration, August 2009. www.osha.gov
	U.S. OSHA Flammable and Combustible Liquids Standard	Flammable and Combustible Liquids, Occupational Safety and Health Standards (29 CFR 1910.106), U.S. Occupational Safety and Health Administration. www.osha.gov
	U.S. DOT PHMSA (Pipeline and Hazardous Materials Safety Administration)	Department of Transportation (DOT) - Pipeline and Hazardous Materials Safety Administration (PHMSA). http://www.phmsa.dot.gov/ (accessed 19-September-2013)
	SEMS - BSEE Safety and Environmental Management Systems for Offshore facilities	The U.S. Bureau of Safety and Environmental Enforcement (BSEE), http://www.bsee.gov/Regulations-and-Guidance/Safety-and-Environmental-Management-Systems---SEMS/Safety-and-Environmental-Management-Systems---SEMS.aspx (accessed 19-September-2013)
	California Accidental Release Prevention Program	California Accidental Release Prevention (CalARP) Program, CCR Title 19, Division 2, Office of Emergency Services, Chapter 4.5, June 28, 2004. www.oes.ca.gov
	Contra Costa County Industrial Safety Ordinance	Contra Costa County Industrial Safety Ordinance. www.co.contra-costa.ca.us
	Delaware Extremely Hazardous Substances Risk Management Act	Extremely Hazardous Substances Risk Management Act, Regulation 1201, Accidental Release Prevention Regulation, Delaware Department of Natural Resources and Environmental Control, March 11, 2006. www.dnrec.delaware.gov
	Nevada Chemical Accident Prevention Program	Chemical Accident Prevention Program (CAPP), Nevada Division of Environmental Protection, NRS 459.380, February 15, 2005. http://ndep.nv.gov/bapc/capp/capp.html
	New Jersey Toxic Catastrophe Prevention Act	Toxic Catastrophe Prevention Act (TCPA), New Jersey Department of Environmental Protection Bureau of Chemical Release Information and Prevention, N.J.A.C. 7:31 Consolidated Rule Document, April 17, 2006. www.nj.gov/dep

Environmental	EPA SARA Title III - U.S. EPA Superfund	U.S. Environmental Protection Agency (EPA), Superfund Amendments and Reauthorization Act (SARA), http://www.epa.gov/superfund/policy/sara.htm (accessed 19-September-2013)
	NPFC - U.S. Coast Guard National Pollution Funds Center	U.S. Coast Guard National Pollution Funds Center (NPFC), http://www.uscg.mil/npfc/laws_and_regulations.asp (accessed 19-September-2013)
Security	DHS - Department of Homeland Security - Facility Vulnerability Assessments (Tiers)	DHS Chemical Security, http://www.dhs.gov/topic/chemical-security (accessed 19-September-2013) and https://www.dhs.gov/critical-infrastructure-vulnerability-assessments (accessed 19-September-2013)
	DHS - U.S. Coast Guard	U.S. Coast Guard, Department of Homeland Security, http://www.uscg.mil/ (accessed 19-September-2013)

Table A 2. International Regulations

Australia Australian National Standard for Control of Major Hazard Facilities	Australian National Standard for the Control of Major Hazard Facilities, NOHSC: 1014, 2002. www.docep.wa.gov.au/
Canada Canadian Environmental Protection Agency, Environmental Emergency Planning	Environmental Emergency Regulations (SOR / 2003-307), Section 200, Environment Canada. www.ec.gc.ca/CEPARegistry/regulations
China China Safety Administration Rules on Dangerous Chemicals	Safety Administration Rules on Dangerous Chemicals; Effective 01-Dec-2011.
China Guidelines for Process Safety Management	Guidelines for Process Safety Management, AQ/T3034-201o; Effective 01-May-2011.
Europe European Commission Seveso II Directive (Note: Seveso III scheduled for 2015)	Control of Major-Accident Hazards Involving Dangerous Substances, European Directive Seveso II (96 / 82 / EC). http://ec.europa.eu/environment/seveso/legislation.htm
Europe European Commission REACH	Registration, Evaluation, Authorisation and Restriction of Chemicals. http://ec.europa.eu/enterprise/sectors/chemicals/reach/index_en.htm. Effective June 1, 2007.
France Ministry of Interior Orsec	Orsec (Organisation de la réponse de sécurité civile). Translated: Organization of the civil protection response. http://www.interieur.gouv.fr/Actualites/Dossiers/Le-plan-Orsec-a-60-ans.
Japan	High Pressure Gas Safety Act See discussion from the High Pressure Gas Safety Institute of Japan: https://www.khk.or.jp/english/faq.html

Korea Korean Occupational Safety and Health Agency, Process Safety Management	Korean Occupational Safety and Health Agency, Industrial Safety and Health Act, Article 20, Preparation of Safety and Health Management Regulations. Korean Ministry of Environment, Framework Plan on Hazards Chemicals Management, 2001-2005. http://english.kosha.or.kr/main
Malaysia Department of Occupation Safety and Health Ministry of Human Resources Malaysia	Malaysia, Department of Occupational Safety and Health (DOSH) Ministry of Human Resources Malaysia, Section 16 of Act 514. http://www.dosh.gov.my/doshV2/
Norway Offshore	See [Khorsandi 2011]
Singapore Singapore Standards Council SS 506 : Part 3 : 2013	Occupational safety and health (OSH) management systems – Part 3 : Requirements for the chemical industry http://www.mom.gov.sg/workplace-safety-health/safety-health-management-systems/Pages/default.aspx
United Kingdom Offshore	See [Khorsandi 2011]
United Kingdom Health and Safety Executive COMAH Regulations	Control of Major Accident Hazards Regulations (COMAH), United Kingdom Health & Safety Executive (HSE), 1999 and 2005. www.hse.gov.uk/comah/

Table A 3. Organizations and Voluntary Industry Standards

ACC - American Chemistry Council Responsible Care © - Management System	American Chemistry Council, 1300 Wilson Blvd., Arlington, VA 22209. http://responsiblecare.americanchemistry.com/Responsible-Care-Program-Elements/Management-System-and-Certification (accessed 18-September-2013)
ACC - American Chemistry Council Responsible Care © - Security Code	ACC 2013 http://responsiblecare.americanchemistry.com/Responsible-Care-Program-Elements/Responsible-Care-Security-Code (accessed 18-September-2013)
ACC - American Chemistry Council Responsible Care © - Process Safety Code	ACC 2013 http://responsiblecare.americanchemistry.com/Responsible-Care-Program-Elements/Process-Safety-Code (accessed 18-September-2013)
ACC - American Chemistry Council Responsible Care © - Performance Metrics	ACC 2013 http://responsiblecare.americanchemistry.com/Responsible-Care-Program-Elements/Performance-Measures-and-Reporting-Guidance (accessed 18-September-2013)

API - American Petroleum Institute Recommended Practices	American Petroleum Institute, 1220 L Street, NW, Washington, D.C., 20005. www.api.org (accessed 18-September-2013)
CIAC Responsible Care® Chemical Industry Association of Canada	http://www.canadianchemistry.ca/index.php/en/index (accessed 09-Mar-2014) Responsible Care® http://www.canadianchemistry.ca/responsible_care/index.php/en/respons ible-care-history (accessed 09-Mar-2014)
ISO 9000 - International Standards Organization Quality management series	ISO Quality Management Series, http://www.iso.org/iso/home/standards/management-standards/iso_9000.htm (accessed 20-September-2013) **Includes the following:** ISO 9001:2008 - Quality management system requirements ISO 9000:2005 - Basic concepts and language ISO 9004:2009 - Improving quality management system efficiency and effectiveness ISO 19011:2011 - Internal and external quality management systems audit guidance
ISO 14000 - International Standards Organization Environmental management series	ISO Quality Management Series, http://www.iso.org/iso/home/standards/management-standards/iso14000.htm (accessed 20-September-2013) **Includes the following:** ISO 14000:2004 - Environmental management system requirements ISO 14004:2004 - Environmental management system guidelines on principles, systems and support techniques ISO 14064-1:2006 - Greenhouse gases -- Part 1: Specification with guidance at the organization level ISO 14006:2011 - Environmental management systems guidelines for incorporating ecodesign
ISO 26000 - International Standards Organization Social Responsibility	http://www.iso.org/sites/iso26000launch/documents.html (accessed 08-April-2014) Includes references to stakeholder involvement and engagement (Clause 5) and the environment (Clause 6)

OHSAS 18000/18001/18002	OHSAS http://www.ohsas-18001-occupational-health-and-safety.com/index.htm (accessed 20-September-2013)
Occupational safety and health Assessment Series	**Incorporates these standards:** BS8800:1996 Guide to occupational safety and health management systems DNV Standard for Certification of Occupational safety and health Management Systems(OHSMS):1997 Technical Report NPR 5001: 1997 Guide to an occupational safety and health management system Draft LRQA SMS 8800 Health & safety management systems assessment criteria SGS & ISMOL ISA 2000:1997 Requirements for Safety and Health Management Systems BVQI SafetyCert: Occupational Safety and Health Management Standard Draft AS/NZ 4801 Occupational safety and health management systems Specification with guidance for use Draft BSI PAS 088 Occupational safety and health management systems UNE 81900 series of pre-standards on the Prevention of occupational risks Draft NSAI`SR 320 Recommendation for an Occupational safety and health (OH and S) Management System

Table A 4. Consensus Codes

ANSI - American National Standards Institute	American National Standards Institute, 25 West 43rd Street, New York, New York, 10036. www.ansi.org
API - American Petroleum Institute	American Petroleum Institute, 1220 L Street, NW, Washington, D.C., 20005. www.api.org
ASME - American Society of Mechanical Engineers	American Society of Mechanical Engineers, Three Park Avenue, New York, New York, 10016. www.asme.org
ISEE - The Instrumentation, Systems and Automation Society / International Electrotechnical Commission	The Instrumentation, Systems, and Automation Society, 67 Alexander Drive, Research Triangle Park, NC 27709. www.isa.org
NFPA - National Fire Protection Association	National Fire Protection Association, 1 Batterymarch Park, Quincy, Massachusetts, 023169. www.nfpa.org

Table A 5. Organizations Committing Efforts to Process Safety

AFPM American Fuel & Petroleum Manufacturers	http://www.afpm.org/ (accessed 09-Mar-2014) Advancing process safety programs http://www.afpm.org/Safety-Programs/ (accessed 09-Mar-2014) http://www.afpm.org/Advancing-Process-Safety-Programs/ (accessed 09-Mar-2014)
API American Petroleum Institute	American Petroleum Institute, 1220 L Street, NW, Washington, D.C., 20005. www.api.org
CCPS Center for Chemical Process Safety Risk Based Process Safety (RBPS)	Guidelines for Risk Based Process Safety, AIChE and John Wiley & Sons, 2007. https://www.aiche.org/ccps (accessed 09-Mar-2014)
Cefic - European Chemical Industry Council Responsible Care©	The European Chemical Industry Council (Cefic), Avenue E. van Nieuwenhuyse, 4 box 1, B-1160 Brussels. www.cefic.org (accessed 18-September-2013)
EMAS - European Union (EU) Eco-Management and Audit Scheme	The EU Eco-Management and Audit Scheme (EMAS) is a management instrument developed by the European Commission for companies and other organisations to evaluate, report, and improve their environmental performance. http://ec.europa.eu/environment/emas/ (accessed 18-September-2013)
ILO International Labor Organisation	Prevention of major industrial accidents http://www.ilo.org/global/publications/ilo-bookstore/order-online/books/WCMS_PUBL_9221071014_EN/lang--en/index.htm (accessed 09-Mar-2014)
OECD The Organisation for Economic Co-operation and Development (OECD)	http://www.oecd.org/ (accessed 09-Mar-2014) Risk management of installations and chemicals http://www.oecd.org/chemicalsafety/risk-management/ (accessed 09-Mar-2014)
PSAP MIT Partnership for a Systems Approach to Safety (PSAP)	http://psas.scripts.mit.edu/home/ (accessed 07-April-2014) A cross-disciplinary effort that recognizes complexity when managing risks. Applies to process safety risk reduction and can facilitate improvements in process safety performance.

APPENDIX B: RECENT ADVANCES IN PROCESS SAFETY METRICS

This appendix provides a brief overview and specific metric-related references detailing recent advances in identifying and selecting metrics affecting process safety performance. The following discussion is based on the references listed in Table B 1.

Figure B 1 conveys an image which can help an organization identify and choose appropriate metrics affecting process safety performance, with the different levels described as follows:

- Corporate or company level indicators - applies to all facilities

    The corporate metrics reflect how the company policies are being used at the facilities.

- Facility level indicators – applies to all process units

    The facility level metrics reflect how the facility's policies are being used across the site.

- Process unit-specific level indicators – applies to all manufacturing units

    The process unit metrics, specific for monitoring process safety performance, reflect how the facility is managing its process safety systems across the site.

Each of these metrics can be selected and monitored to help ensure that an organization's operating risks meet the ALARP criteria [Baybutt 2014] and that the process safety systems (the pillars and associated elements of a process safety management program) have been implemented and are being maintained [CCPS 2007a, Sepeda 2010].

The metrics selected for monitoring the process safety performance must be controlled by the level of the organization expected to monitor and respond to it. Hence, there are metrics specific to a manufacturing unit, such as the actual loss of wall thickness of a reactor exposed to corrosive conditions, which may not be suitable for tracking at the corporate level. For this metric example, a useful facility metric could involve the activity of performing the proper non-destructive test (NDT) and inspection, part of the reactor's preventive maintenance (PM) program. The facility ensures inspection quality (in advance) by selecting a certified inspector. A useful company metric could identify whether the facility's PMs are being performed per the required PM schedule. And, if the NDT inspection criteria are not met (i.e., the reactor wall thickness is too thin), then those operating the manufacturing unit can respond to the failure of the NDT by issuing a capital project to replace the reactor.

Table B 1. Some Recent Advances in Process Safety Metrics

API	2010	American Petroleum Institute (API), Process Safety Performance Indicators for the Refining and Petrochemical Industries, RP 754, 1st Edition, April 2010.
Azizi	2013	Azizi, W, "Process Safety: how do you measure up?," the chemical engineer (tce), The Institution of Chemical Engineers (IChemE), Rugby, Warwickshire, CV21, 3HQ, UK, August, 2013, pp. 31-34. www.tcetoday.com (accessed 16-September-2013)
CCPS	1996	Center for Chemical Process Safety (CCPS), Guidelines for Integration of PSM, ES&H and Quality, AIChE, New York, 1996. (Note: this Guideline is the update of the 1996 book).
CCPS	2007	Center for Chemical Process Safety (CCPS), Guidelines for Risk Based Process Safety, AIChE and John Wiley & Sons, Inc., Hoboken, New Jersey, 2007.
CCPS	2010	Center for Chemical Process Safety (CCPS), Guidelines for Process Safety Metrics, AIChE and John Wiley & Sons, Inc., Hoboken, New Jersey, 2010.
CCPS	2011	Center for Chemical Process Safety (CCPS), Process Safety Leading and Lagging Metrics, Revised: January 2011. [CCPS 2011b]
CCPS	2011	Center for Chemical Process Safety (CCPS), Conduct of Operations and Operational Discipline, AIChE and John Wiley & Sons, Inc. Hoboken, NJ, 2011. [CCPS 2011c]
CCPS	2012	Center for Chemical Process Safety (CCPS), Recognizing Catastrophic Incident Warning Signs, AIChE and John Wiley & Sons, Inc., Hoboken, NJ, 2012.
CEFIC	2011	CEFIC, Guidance on Process Safety Performance Indicators, 2nd Edition, May 2011.
Hopkins	2009	Hopkins, A., "Thinking about process safety indicators," Safety Science, Elsevier, 47 (2009) 460-465.
HSE	2006	Health and Safety Executive (HSE), Developing process safety indicators, A step-by-step guide for chemical and major hazard industries, HSG 254, 2006.
ISO	2008	International Standards Organization, The integrated use of management system standards, Edition 1, 2008. http://www.iso.org/iso/home/store/publications_and_e-products/publication_item.htm?pid=PUB100068 (accessed 20-September-2013)
Klein	2011	Klein, J. A., and B. K. Vaughen, "Implementing an Operational Discipline Program to Improve Plant Process Safety," AIChE Chemical Engineering Progress (CEP), June 2011, pp. 48-52.
OECD	2008	Organisation for Economic Co-operation & Development (OECD), Guidance on Developing Safety Performance Indicators related to Chemical Accident Prevention, Preparedness and Response, Paris, 2008.
OGP	2011	International Association of Oil & Gas Producers (OGP), Process Safety - Recommended Practice on Key Performance Indicators, Report No. 456, November 2011.
Overton	2008	Overton, T. and S. Berger, "Process Safety: How are you doing?," Chemical Engineering Progress (CEP), AIChE, May 2008, pp. 40-43.
Pilkington	2013	Pilkington, G., "Beyond boots and goggles," the chemical engineer (tce), The Institution of Chemical Engineers (IChemE), Rugby, Warwickshire, CV21, 3HQ, UK, August, 2013, pp. 37-38. www.tcetoday.com (accessed 16-September-2013)
Vaughen	2012	Vaughen, B. K., and J. A. Klein, "What you don't manage will leak: A tribute to Trevor Kletz," Process Safety and Environmental Protection (PSEP), Vol. 90, No. 5, September 2012, pp. 411-418.

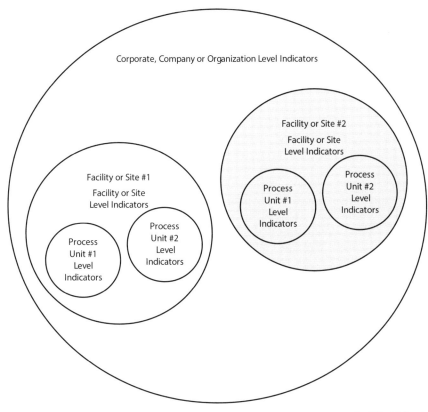

(Similar approach to the types of levels described in HSE's HSG254)

Figure B 1. Identifying and Choosing Appropriate Process Safety Metrics

There is a distinction between the types of metrics, as well. T here are "leading indicators" and "lagging indicators" which help define the type of metric that can be used. The leading indicators reflect the activities of the process units and facilities that are expected by the company and regulators to maintain safe and reliable operations. T hese activities are proactive indicators that reflect implementation of the process safety systems designed for safe and reliable operations. T he lagging indicators reflect the consequences resulting when an organization has not adequately designed or implemented some or all of the essential elements of its process safety program.

For example, a facility could monitor the number of PHAs that are on the schedule, with its leading indicator monitoring the chartering of a P HA Team, including its preparation time before the PHA Team sessions begin (e.g., a few weeks beforehand), whereas lagging indicators could include the number of overdue PHAs (from a scheduling perspective) or incidents which resulted from lack of hazards understanding and proper hazards analysis reviews of these hazards and risks (from the PHA execution perspective). N ote that lagging indicators include loss of containment events (spills) that may or may not have led to toxic releases, fires and explosions causing fatalities, injuries and environmental harm.

A recent insight on choosing and responding to metrics which affect process safety performance can be represented with the diagram presented in Figure B 2, where the "fuzzy" range between leading and lagging indicators is depicted as a continuum, not as a discreet break. This image is based on recent publications and articles that use either the barrier approach or the tiered/pinnacle approach, with the best approach including understanding and monitoring of both personal and organizational operational discipline across all groups (see references in Table B 1). T he best approach combines both the tangible, easy to measure technical metrics and the soft, difficult to measure management system and behavioral metrics. N ote that these approaches are "event driven" approaches, which first identify the essential barriers or layers of protection which help reduce the overall operational risks (based on the worst case event), and then determine which leading and lagging indicators need to be monitored and tracked for each barrier. A dditional discussion on the event driven/barrier approach is described with the Bow Tie discussion in Chapter 3).

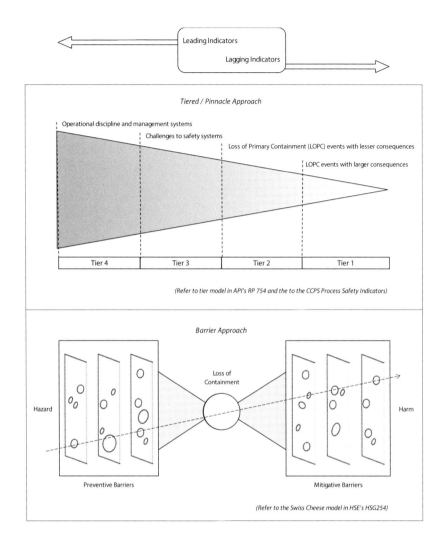

Figure B 2. Range of Process Safety-related Performance Indicators (metrics) for Monitoring Process Safety Systems

APPENDIX C: POTENTIAL ANSWERS DESCRIBING THE NEED FOR SECURING SUPPORT

Potential answers for stakeholders addressing the questions posed in Chapter 2, Section 2.1, are listed in this appendix. These answers should help everyone understand how the SHEQ&S program will be a part of the organization's overall management system, eventually becoming part of the normal work process in the organization. The answers to these questions are as follows:

- Who benefits from the SHEQ&S program?

    Answer: "Everyone benefits from the SHEQ&S program." From those at process unit level to those at the senior management level and those external to the company, such as those living in the surrounding communities or relying on the company's regulatory compliance.

- What are the benefits of the SHEQ&S program?

    Answer: "The benefits of the SHEQ&S program include less work for everyone managing process safety and helping, in part, to sustain a company's survival." Although organizations struggle to do things the right way, let alone multiple ways, everyone from the process unit level to the senior management level benefits from an effectively integrated management system.

- What will the final SHEQ&S program look like?

    Answer: "This is a preliminary plan that provides the vision and goals for the SHEQ&S program. The final program needs to be developed by the program integration team." Please refer to Section 2.3, Figure 2-5, for an image that can be tailored and used, showing how the existing management systems are integrated into the SHEQ&S program.

- How does the SHEQ&S program differ from the current systems?

    Answer: "The SHEQ&S program will not replace existing SHEQ&S management systems. It will take advantage of the existing management systems by eliminating duplicate efforts across groups using common metrics that affect process safety performance."

- How will the change be achieved?

  Answer: "The change will be achieved through a team effort that is supported by everyone in the organization." Approaches on how changes can be achieved are the subject of Chapter 2.

APPENDIX D: DETAILED CASE STUDY FOR DESIGNING AND IMPLEMENTING A SHEQ&S PROGRAM

The Case for a SHEQ&S Program

This appendix illustrates the aspects of a process safety incident to present a case for the SHEQ&S program designed to integrate the different SHEQ&S management systems. The outline for this case study is as follows:

1. The scenario and a description of the process unit process
2. Two adverse impacts: Part 1) The immediate impact on operations
3. Two adverse impacts: Part 2) The delayed impact on the operating risk
4. Additional reflections on the case for a SHEQ&S program

It is hoped that some aspects of this scenario help the reader reflect upon their own experience in a w ay which reinforces the need for an effective and SHEQ&S program.

D-1. The scenario and a description of the process unit process

The incident case study occurs during an economic market collapse, in a process unit used to polymerize a flammable monomer. With the market crash, customers stop ordering polymer. L ack of sales causes the profits to decline and the company responds by curtailing production, making drastic budget cuts across all groups. The company recognizes that these cuts do not guarantee longevity but do improve the odds that the process unit may be able to survive the economic downturn.

The process flow diagram for the polymerization part of the process unit is shown in Figure D 1 where the monomer is polymerized, the unreacted monomer is recovered, and the polymer is processed downstream [US CSB 2011a]. The vinyl fluoride (VF) monomer is handled like liquid natural gas, cooled under pressure, and pumped to the reactor as a liquid. Since the polymerization reaction does not consume all of the VF fed to the reactor, the unreacted VF is separated from the polymer slurry and recycled. T he slurry is then pumped to the Slurry Tanks for subsequent filtering, drying and storage before use. The VF monomer vaporizes at atmospheric temperature and pressure, and will form an explosive mixture when exposed to air.

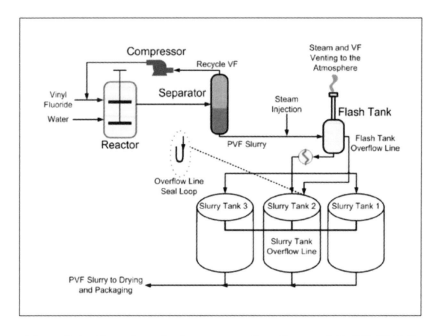

Figure D 1. The polymerization process flow diagram [US CSB 2011a]

One group adversely affected by the budget cuts was the maintenance and operations group, with some of their mechanics, electricians and operators laid off. However, these experienced staff members were essential for effectively managing the process safety risks during normal operations. One way to look at the normal operations risk management is on the risk matrix shown in Figure D 2. With proper understanding of the unmitigated risk combined with proper design and implementation of the process safety systems, the process unit's normal operating risks had been relatively low. The process unit's day-to-day residual risk had been reduced to the tolerable risk level. By operating and maintaining with process safety programs and systems for decades, the process unit's resource allocation across the SHEQ&S groups were optimized, with an appropriate balance of its resources to minimize its operational risk. However, the operational risk increased from the reduction in resources, with the shift on the overall risk curve moving to the left from normal operations (1) to the stressed operations (2), as is shown in Figure D 3.

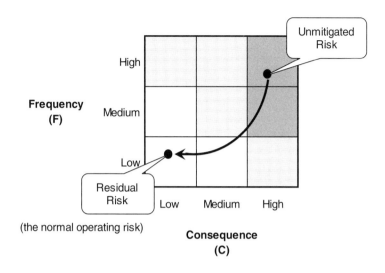

Figure D 2. The residual risk where the process unit normally operates

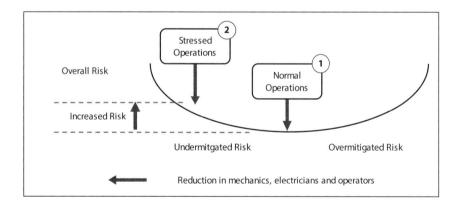

Figure D 3. The increase in overall operational risk with the staffing reduction

It is important to recognize that there are many process safety systems that must be supported during normal operations to help manage the risk. T hese process safety systems include identification of the hazards during all phases of operation (startup, normal, shutdown and emergencies), understanding of the normal process and equipment design intents, thorough hazards and risk analyses, standardized operating procedures, facility-wide safe work practices, and equipment integrity tests and inspections. The process safety systems inherent in normal operations are shown in Table D 1. The unmitigated risk, noted as "high" for the reactor, separator, and compressor is mitigated to "low" after implementation of the equipment-specific independent engineering and administrative layers of protection. The benefits of an SHEQ&S program will become apparent with better communications between the SHEQ&S groups when tough decisions must be made across the board.

Table D 1. A process safety risk evaluation reflecting normal operations

		Reactor	Separator	Compressor	Flash Tank	PVF Slurry Tanks
Normal Operations						
	Vinyl Fluoride present?	Yes Polymerized at high pressure to PVF	Yes	Yes	Not significant	Not significant
	Vinyl Fluoride Hazard	Above LEL	Above LEL	Above LEL	Vented to atmosphere	If any residual, below LEL
	Polyvinyl Floride present?	Yes	Yes	Yes	Yes	Yes Slurry
	Unmitigated Risk	High	High	High	Medium	Low
Process Safety Systems						
Barriers						
Process Technology and Process Hazards Analysis	Equipment Design Hazards and risk analyses	Yes Pressure vessel	Yes	Yes	Yes	Yes
Operating Procedures	SOLs based on equipment design	Yes	Yes	Yes	Yes	Yes
Safe Work Practices	Hot Work Permit	Yes (applies to facility)				
Equipment Integrity	Scheduled tests and inspections	Yes Pressure vessel	Yes	Yes	Yes	As needed
	Residual (mitigated) Risk	Low	Low	Low	Low	Low

D-2. Two adverse impacts: Part 1) The immediate impact on operations

The immediate impact on the operations was the drastic reduction in resources needed to run the process unit. The cost cuts included labor and capital expenditures and were distributed across the SHEQ&S groups. The reduction in the SHEQ&S group resource allocations between the normal operations (Scenario 1) and the curtailed operations (Scenario 2) is shown for the process unit in Figure D 4.

Although not apparent in the "bottom line" metric being measured, the equipment life cycle and programs associated with ensuring safe and reliable operations were also affected by the cost cuts. T he decision makers at the corporate level impacted all equipment stages, delaying all capital project expenditures and reducing the staff working on continuous improvement efforts at the facilities. The decision makers at the facility level had to listen to corporate, stopping the minor capital projects, including those managing changes to the process unit's equipment, and reducing staff, as well. At the polymerization process unit level, the idled production equipment's preventive maintenance test and inspections were either delayed or simply not performed at all. ("Why maintain something that is not running?") A summary of the effects of the cost-cutting decisions on the different stages of the equipment life shown at the different levels in the organization is shown in Figure D 5.

It is important to recognize at this point that few of the decision makers considered the long term view at all since the future of the process unit was unknown. All believed that the process unit would have no future at all if it didn't make the cuts immediately. The effects of the different decision makers, whether they are situated at the corporate, facility or process unit level, on all stages of the equipment's life cycle are compared and shown versus the stages in Figure D 5, as well. E very decision maker affects every stage to some degree. Corporate decision makers should not expect process units to operate as long as designed if essential maintenance is not performed. The equipment will fail if it is operated without maintenance.

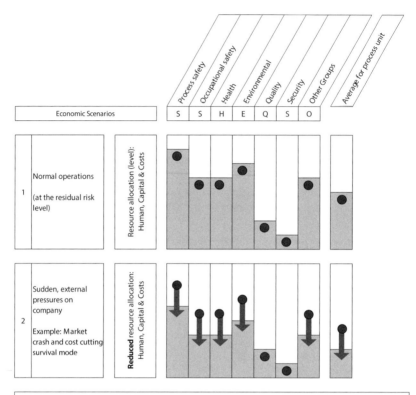

Figure D 4. The impact on the SHEQ&S group resource levels

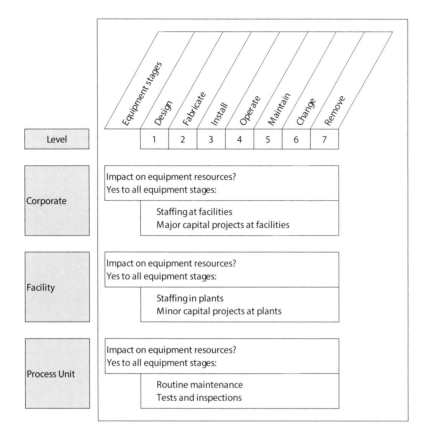

Figure D 5. The impact from resource-related decisions on the equipment's integrity and reliability stages

D-3. Two adverse impacts: Part 2) The delayed impact on risk

This case study continues by showing how the process safety-related risk increased from its low, tolerable level to a high, unacceptable level due to the delayed or eliminated preventive maintenance tasks. When demand increased again, the process unit resumed its operations but did not have time to increase its staff which had been reduced during the layoffs. Unfortunately, following the pressure to satisfy the market, operations resumed without addressing the preventive maintenance tests and inspections on the compressor and the slurry tanks that needed to be addressed during the idle time. In particular, inspections performed months before the economic slowdown had found corroded vapor overflow lines between slurry tanks.

The consequences of the missed preventive maintenance efforts were not observed until months later, shortly after the process unit finished its annual shutdown. When the compressor failed during the restart operation, the separator could not operate at its expected design limits, resulting in more monomer vapors being retained in the PVF slurry sent to the slurry tanks. The monomer subsequently vaporized and travelled from the slurry tanks in service through the corroded and unrepaired overflow lines to the locked out / out-of-service slurry tank. These vapors accumulated to a level above the LEL in the vapor space of the out-of-service slurry tank. Although everyone thought the delayed hot work maintenance from the shutdown could be performed safely on the out-of-service slurry tank, the accumulated vapors ignited and resulted in a fatality. As is shown in Figure D 6, the actual risk once operations resumed was no longer in its tolerable, low state. The change in safe risk management is evident when comparing the "Barrier Integrity" question responses during normal operations (Table D 1) to the responses once operations resumed (Table D 2).

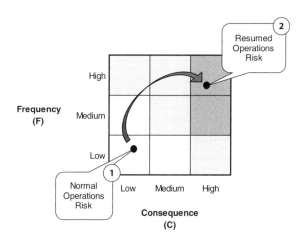

Figure D 6. The actual risk once operations resumed

Table D 2. The process safety risk evaluation once operations resumed

		Reactor	Separator	Compressor	Flash Tank	PVF Slurry Tanks
Resumed Operations after resource reduction						
	Vinyl Fluoride present?	Yes	Yes	Yes	Yes	Yes (unknown)
	Vinyl Fluoride Hazard	Above LEL	Above LEL	Above LEL	Vented to atmosphere	Above LEL
	Polyvinyl Floride present?	Yes	Yes	Yes	Yes	Yes Slurry
	Unmitigated Risk	High	High	High	Medium	High
	Process Safety Systems					
	Barrier Integrity?					
Operating Procedures	Operated within SOLs?	Yes	No	No	Yes	Yes
Safe Work Practices	How Work Permit?	Yes (applies to facility)				
Equipment Integrity	Scheduled tests and inspections?	Yes Pressure vessel	No	No	No	No
	Actual Risk	Low	High	High	High	High

Although hindsight provides the investigation team with the "what should have been done" recommendations, it is important to recognize that everyone making decisions at the time the incident occurred were making them on the best available information they had at the time [Vaughen 2011]. If the risk allocation image is revisited, the reduced resource allocation image shown in Figure D 4 can be transformed as a company risk image showing the increase in the SHEQ&S risks between normal operations and the resumed operations in Figure D 7. During normal operations, all groups are operating at the company's tolerable risk level (low). However, the budget cut induced stressed operations that resulted in process safety risks which were inadequately addressed after operations resumed. The company's risk (its exposure) was assumed to still be at an acceptable (low) level because no information was readily available to highlight the gaps which had developed. In reality, the risk for the company was unknowingly above its tolerable risk, a situation which could have been identified by an integrated SHEQ&S system.

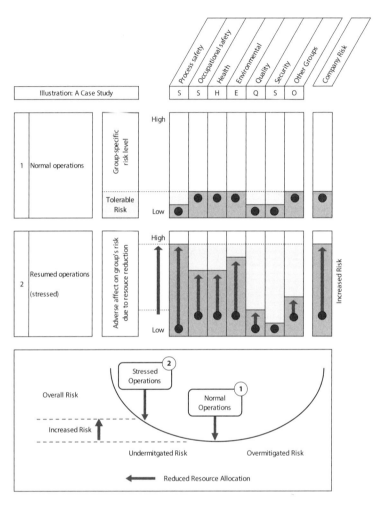

Figure D 7. The increase in the company risk when stressed

D-4. Additional reflections on the case for an SHEQ&S program

If monitoring metrics that affect process safety performance were present in an effectively designed and implemented SHEQ&S system, the impacts on the decisions made in one group and their potential impact on other groups could have been better shared and communicated across the company. Ensuring the safe startup of processes over the life of a facility is one of the elements in the RBPS pillar of managing risk [CCPS 2007a, Sepeda 2010]. Without a proper operational readiness review, incidents occur. In addition, without a proper organizational change review, incidents occur. Since other references address operational readiness (including pre-start up reviews) and managing organizational change [CCPS 2007a, CCPS 2007b, CCPS 2013], they will not be discussed here.

Although there are many process safety-related metrics from which to choose, a leading indicator option for this case study could be monitoring and tracking the number of scheduled preventive maintenance tests and inspections on critical equipment. The process unit's delayed or missing equipment tests and inspection would have been evident and could have been raised as an issue during reviews. Proactive actions resulting from this gap would have included re-scheduling and performing the tests and inspections, with proper responses, as needed. Knowledge is a key element for good decisions. It is hoped that future incidents which may cause fatalities, injuries, environmental harm and property destruction can be avoided with shared knowledge through an effective SHEQ&S program.

In the context of maintaining the equipment's integrity across its life cycle, poor equipment reliability and equipment performance will combine to reduce the production rates and may adversely affect product quality, as well. The maintenance department's objectives include economically and effectively preserving physical assets while safely providing continuous availability of the process equipment. It is also necessary to recognize that the preventive tests and inspections are designed to extend the equipment's useful life, including detecting and responding to equipment conditions that will require costly repairs if not addressed early when the issues are relatively small. In essence, routine preventive maintenance extends the equipment's useful life and helps avoid an untimely failure. Tools used to support maintenance that would provide direction to help identify useful metrics include operating procedures, work orders, data and information entry and retrieval, priority settings, tracking of current and future work, all of which can provide useful metrics for tracking within an SHEQ&S program.

As shown in this case study, an effectively designed and implemented SHEQ&S program should help to proactively identify process safety gaps, allowing time to address and correct issues before it's too late.

APPENDIX E: EQUIPMENT INTEGRITY IN THE EQUIPMENT LIFE CYCLE

The equipment integrity must be maintained during its useful life cycle, from the equipment's design to the end of its useful life and ultimate removal. The equipment life cycle is represented with seven distinct stages shown in Figure E 1: design, fabricate, install, operate, maintain, change and remove.

People at every level in the organization directly or indirectly affect the equipment's integrity at some point during each stage:

- Engineering design must address the hazards of the processes and the materials.

- Construction (includes procurement) must fabricate and install the equipment according to the design specifications.

- Operations must operate the equipment within its safe operating limits.

- Maintenance must perform the equipment's preventive maintenance tests and inspections to extend the useful life of the equipment, and depending on the results, must initiate a response to address gaps based on the test or inspection.

And, most importantly, everyone must help manage the equipment's changes through all of its stages from its design to the end of its useful life.

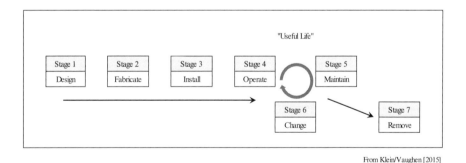

From Klein/Vaughen [2015]

Figure E 1. The stages in equipment's life – the equipment life cycle

The equipment is a part of the physical assets which are used to meet the company's objectives: adding value to incoming materials by transforming them into products. The RBPS process safety systems that are used to help manage the equipment integrity risks include: Compliance with Standards, Process Knowledge and Management, and Hazard Identification and Risk Analysis (Stage 1); Contractor Management (Stages 2 and 3); Operating Procedures and Safe Work Practices (Stage 4); Asset Integrity and Reliability (Stage 5); and Management of Change and Operational Readiness (Stages 1, 2, 3, 4, 5, 6 and 7) [CCPS 2007a, Sepeda 2010]. When a comprehensive process safety program addresses all stages in the equipment life cycle, the equipment will not suddenly fail and lead to a process safety incident.

APPENDIX F: THE SHEQ&S MANAGEMENT SYSTEM MAPPING SURVEY

The example management systems mapping survey described in this appendix has been pre-populated with the U.S. OSHA PSM and U.S. EPA RMP regulatory expectations. T he "Other" column, if needed, can be populated by each organization and facility with other regulations, standards or guidelines that apply in other jurisdictions such as: the Canadian EPA Environmental Emergency Regulations and the European Directive Seveso II.

This survey is framed using the Risk Based Process Safety (RBPS) management system guidance provided by the CCPS [CCPS 2007a, Sepeda 2010]. As is shown Figure F-1, there are twenty elements identified for a s uccessful management system based on the following four pillars:

1) Commit to process safety
2) Understand the hazards and risks
3) Manage risk
4) Learn from experience.

The survey is designed to evaluate the management systems across each of the SHEQ&S groups at both corporate and facility levels in the organization. I n particular, gaps in the management systems identified in these surveys must be clearly understood and addressed, as needed.

The premise of this survey is to-
Successfully reduce the work demands on the different SHEQ&S groups by understanding and enhancing the existing management systems, not creating new work processes.

The questions posed in the SHEQ&S system mapping survey focus on the systems used to manage an organization's operational risk across the SHEQ&S groups. Since global organizations have facilities under different jurisdictions and regulations, its corporate standards and guidelines must be performance based, allowing each facility to develop their prescriptive, facility-specific standards and guidelines. For this reason, there are two surveys tailored to address the corporate "performance" based standards and the facility "prescriptive" based standards.

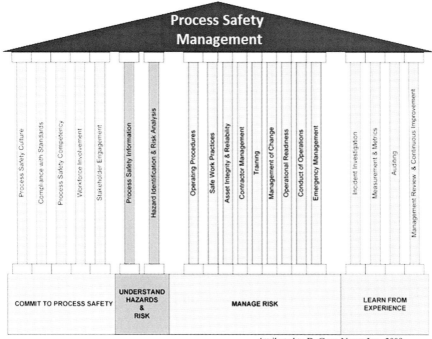

Attributed to D. Guss, Nexen Inc., 2009.
Resourced by the CCPS [Sepeda 2010].

Figure F-1. The CCPS Process Safety Management System

The "SHEQ&S Mapping System" workbook contains two surveys, the first with corporate-level management questions and the second with facility-level management questions. Although these surveys are included and described in this appendix, please access the CCPS website for the latest version of these surveys at:

http://www.aiche.org/ccps/publications/metrics-tools

Description of the Corporate Management Level Survey

Goal: to identify gaps between the corporate process safety standards and guidelines and the process safety-related regulatory expectations.

Figure F-2 describes the line-by-line mapping (comparison) of the corporate standards, guidelines and policies to the CCPS RBPS pillars/elements (on the left) and applicable regulations (on the right). Note that the example in this template is pre-populated with the process safety-related elements from the U.S. OSHA PSM and U.S. EPA RMP. SHEQ&S system gaps, if any, can be noted in the survey with a corporate-level action to address the gap.

Figure F-2. Rows and columns in the corporate SHEQ&S management systems mapping survey ("Appendix F, Table F-1")

Description of the Facility Management Level Survey

Goal: to identify gaps between the facility standards and guidelines and both the corporate process safety standards and the process safety-related regulatory expectations.

Figure F-3 describes the line-by-line mapping (comparison) of the facility standards, guidelines, and policies to the corporate standards, guidelines, and policies (on the left), to the CCPS RBPS pillars/elements (on the left) and to other applicable regulations (on the right). Note that this example in this template is prepopulated with the process safety-related elements from the U.S. OSHA PSM and U.S. EPA RMP. S HEQ&S system gaps, if any, can be noted in the survey with a facility-level action to address the gap.

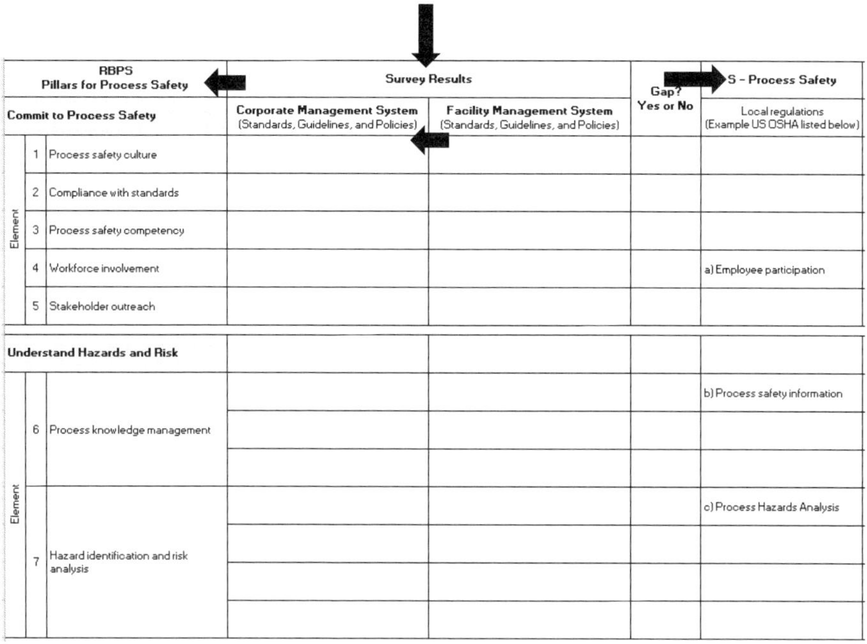

RBPS Pillars for Process Safety		Survey Results		Gap? Yes or No	S – Process Safety
Commit to Process Safety		Corporate Management System (Standards, Guidelines, and Policies)	Facility Management System (Standards, Guidelines, and Policies)		Local regulations (Example US OSHA listed below)
1	Process safety culture				
2	Compliance with standards				
3	Process safety competency				
4	Workforce involvement				a) Employee participation
5	Stakeholder outreach				
Understand Hazards and Risk					
6	Process knowledge management				b) Process safety information
7	Hazard identification and risk analysis				c) Process Hazards Analysis

Figure F-3. Rows and columns in the facility SHEQ&S management systems mapping survey ("Appendix F, Table F-2")

The following pages provide a sample of the Appendix F SHEQ&S Management System Mapping Tool at the time of publication. Please access the CCPS website for the latest version of these surveys at:

http://www.aiche.org/ccps/publications/metrics-tools

Table of Contents

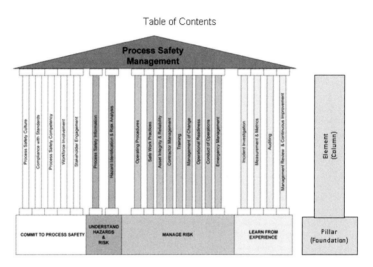

Attributed to D. Guss, Nexen Inc., 2009.
Resourced by the CCPS, from Sepeda, A. L., "Understanding Process Safety Management," Chemical Engineering Progress, August 2010, Vol. 106, No. 6, pp. 26-33.

Worksheet

1)	Table of Contents
2)	Corporate Survey - Maps corporate standards versus regulatory expectations
3)	Facility Survey - Maps facility standards versus corporate and regulatory expectations

Figure F-4. Table of Contents for Appendix F "Surveys"

The CCPS Risk Based Process Safety (RBPS) SHEQ&S Management System Mapping Tool - Corporate Systems

RBPS Pillars for Process Safety	Survey Results — Corporate Management System (Standards, Guidelines, and Policies)	Gap? Yes or No	S - Process Safety			H - Occupational Health and Safety			E - Environmental			Q - Quality		S - Security	
			US OSHA PSM Elements	Other		US OSHA	Other		US EPA RMP Parts	Other		Corporate Certifications		US DHS	Other
Commit to Process Safety															
1 Process safety culture															
2 Compliance with standards															
3 Process safety competency															
4 Workforce involvement			a Employee participation						63 Employee participation						
5 Stakeholder outreach									see RMP Subpart E requirements						
Understand Hazards and Risk															
6 Process knowledge management			b) Process safety information						65 Process safety information						
									10 (a)Threshold quantity analysis						
									48 Safety information						
7 Hazard identification and risk analysis			c Process Hazard Analysis						67 Process hazard Analysis						
									25 Worst-case release scenario analysis						
									33 Off-site impacts analysis - alternative						

Figure F-5. Example of Corporate Survey for Appendix F

Figure F-6. Example of Facility Survey for Appendix F

APPENDIX G: THE PROCESS SAFETY PERSONNEL COMPETENCY SURVEY

The example personnel competency survey provided in this appendix has been pre-populated the U.S. OSHA PSM and U.S. EPA RMP regulatory expectations. The "Other" column, if needed, can be populated by each organization and facility with other regulations, standards or guidelines that apply in other jurisdictions, such as: the Canadian EPA Environmental Emergency Regulations and the European Directive Seveso II.

The premise of this survey is to-
   Successfully implement the integrated SHEQ&S management system with competent personnel across all levels in an organization.

The questions posed in the process safety competency surveys focus on the personnel applying the corporate and facility's process safety-specific management systems. Gaps in personnel accountability, if any, are identified quickly, helping ensure that everyone knows what their role is, from those responsible for providing the resources to execute the corporate or facility programs to those responsible for executing the design, construction, operation or maintenance of the equipment in the field.

This survey is framed using the Risk Based Process Safety (RBPS) management system guidance provided by the CCPS [CCPS 2007a, Sepeda 2010]. As is shown Figure G-1, there are twenty elements identified for a successful management system based on the following pillars:

1) Commit to process safety
2) Understand the hazards and risks
3) Manage risk
4) Learn from experience.

The survey is designed to evaluate the competency of personnel responsible for resourcing or executing process safety-related systems across each of the SHEQ&S groups at all levels in the organization (corporate, facility and process unit, see the organization level terminology provided in Table 2-1). In particular, gaps in the personnel competencies that are identified with these surveys must be clearly understood and addressed, as needed.

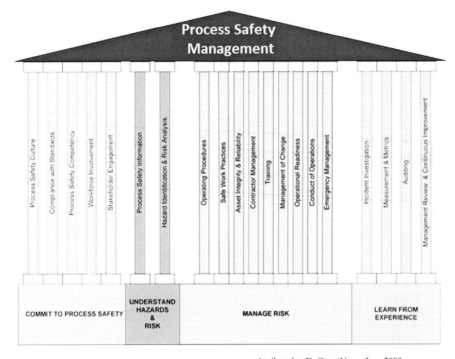

Attributed to D. Guss, Nexen Inc., 2009.
Resourced by the CCPS [Sepeda 2010].

Figure G-1. The CCPS Process Safety Management System

The following pages provide samples of the personnel competency surveys provided in Appendix G at the time of publication. The examples reflect part of the "Senior Leadership" survey based on the separate pillars with corresponding surveys tailored for the "Facility Leadership" and the "Process Unit Leadership" in corresponding worksheets (see the survey list in Table G-1). Please access the CCPS website for the latest version of these surveys at:

http://www.aiche.org/ccps/publications/metrics-tools

The workbook contains the personnel competency surveys across the corporate, the facility and the process unit leadership levels (see the organization level terminology provided in Table 2-1). T he personnel competency survey worksheets are briefly described and listed in Table G-1. The surveys are distributed across the three levels as follows:

Surveys / Pillars   Organizational Leadership Level

2.1, 3.1, 4.1, and 5.1   Senior/Corporate (examples presented below)

2.2, 3.2, 4.2, and 5.2   Facility

2.3, 3.3, 4.3, and 5.3   Process Unit

Table G-1. Copy of the Table of Contents for the Personnel Competency Surveys (Table G-1 in the Excel file for Appendix G)

Worksheet			Worksheet Overview
Table G-1	Table of Contents		Appendix G Table of Contents
Table G-2	RBPS Survey Framework		Shows RBPS framework for Survey worksheets
Table G-3	Survey Level Definitions (Corporate/Senior, Facility, and Process Unit)		Three levels for questions due to different responsibilities
Table G-4	RBPS Pillar Descriptions		From CCPS website
Table G-5	CCPS References		List of CCPS Guidelines for additional RBPS support
Survey 1.1)	Process Safety Accountability		Helps identify gaps in responsibilities
Survey 2.1)	Senior Level	Pillar Commit to Process Safety	Helps identify gaps pertaining to specific elements for this pillar across the three levels
Survey 2.2)	Facility Level		
Survey 2.3)	Process Unit Level		
Survey 3.1)	Senior Level	Pillar Understand Hazards and Risks	Helps identify gaps pertaining to specific elements for this pillar across the three levels
Survey 3.2)	Facility Level		
Survey 3.3)	Process Unit Level		
Survey 4.1)	Senior Level	Pillar Manage Risks	Helps identify gaps pertaining to specific elements for this pillar across the three levels
Survey 4.2)	Facility Level		
Survey 4.3)	Process Unit Level		
Survey 5.1)	Senior Level	Pillar Learn from Experience	Helps identify gaps pertaining to specific elements for this pillar across the three levels
Survey 5.2)	Facility Level		
Survey 5.3)	Process Unit Level		

Table G-2. Copy of the Framework for the Personnel Competency Surveys (Table G-2 in the Excel file for Appendix G)

Table G-2

Framework for the Process Safety Personnel Competency Survey

Basis: Risk Based Process Safety (RBPS) AIChE/CCPS

Survey Question Set				Workbook Tabs		Survey Focus
1) Process safety accountability Who is accountable and where does it apply				1.1	Process Safety Survey	Responsibilities and Applicability
2) Commit to Process Safety				2.1	Senior Level	Responsibilities
				2.2	Facility Level	Responsibilities
				2.3	Process Unit Level	Responsibilities
	Pillar	Element	1	Process safety culture	See CCPS Guidelines **Table G-5** in this appendix	
			2	Compliance with standards		
			3	Process safety competency		
			4	Workforce involvement		
			5	Stakeholder outreach		
3) Understand Hazards and Risk				3.1	Senior Level	Resourcing
				3.2	Facility Level	Implementation
				3.3	Process Unit Level	Resourcing
	Pillar	Elmnt.	6	Process knowledge management	See CCPS Guidelines **Table G-5** in this appendix	
			7	Hazard identification and risk analysis		
4) Manage Risk				4.1	Senior Level	Resourcing
				4.2	Facility Level	Implementation
				4.3	Process Unit Level	Implementation
	Pillar	Element	8	Operating procedures	See CCPS Guidelines **Table G-5** in this appendix	
			9	Safe work practices		
			10	Asset integrity and reliability		
			11	Contractor management		
			12	Training and performance assurance		
			13	Management of change		
			14	Operational readiness		
			15	Conduct of operations		
			16	Emergency management		
5) Learn from Experience				5.1	Senior Level	Implementation
				5.2	Facility Level	Implementation
				5.3	Process Unit Level	Implementation
	Pillar	Element	17	Incident investigation	See CCPS Guidelines **Table G-5** in this appendix	
			18	Measurement and metrics		
			19	Auditing		
			20	Management review and continuous improvement		

Table G-3. Copy of Leadership Definitions for the Personnel Competency Surveys (Table G-3 in the Excel file for Appendix G)

Table G-3				
Definitions for the Survey Levels				
Survey Based on Organizational Level				Terms that may be used in an organizational chart (Refer to **Table 2-1**)
Corporate Level	Senior Level Survey	Other terms for this level: Enterprise Organization	Groups noted in this guideline	Process and Occupational Safety (S), Occupational Health (H), Environmental (E), Quality (Q) and Security (S)
			Staff terminology	Includes president, vice president, executive, chief operating officer (COO), global director, global manager; includes global Process Safety Management (PSM) directors
			Regions	Includes Europe, North America, South America, Asia Pacific, Africa, Middle East
			Competency Centers	Includes process safety management (PSM), environmental, health and safety (EHS), engineering, maintenance, procurement, information services, supply chain, operations, operational excellence, research and development (R&D), sustainability
			Departments or Divisions	Includes financial, legal, taxes, insurance (loss prevention; property and casualty), strategic planning, communications, government relations, auditing, human resources, investor relations Divisions also noted with product-related groupings (e.g., chemicals, refining, upstream, downstream, etc.)
Business Level	Other terms for this level: Business Unit Business Stream Segments			A "business" is typically based on similar technologies or markets, such as refining, chemicals, specialty chemicals, advanced materials, biological, plant sciences, explosives, etc. Business Units may have facilities at different locations across the world
Facility Level	Facility Level Survey	Other terms for this level: Plant Site	Groups noted in this guideline	Process and Occupational Safety (S), Occupational Health (H), Environmental (E), Quality (Q) and Security (S)
			Staff terminology	Includes facility manager, senior managers, assistant managers, deputy managers, engineers, officers; includes facility (site) PSM element owners
			Department terminology	Includes production, operations, maintenance, engineering, projects, quality control and assurance, information technology (IT), raw materials storage and/or warehouse, purchasing, customer service, human resources, administration, accounting, finance
Process Unit Level	Process Unit Level Survey	Other terms for this level: Assets	Staff terminology	Includes operators, mechanics, electricians, technicians, process support engineers, laboratory technician, attendants, workers, line supervisor; includes local PSM element owners
			Hazardous process terminology	Processes that handle hazardous materials and energies with the potential for harm to people, the environment and property if the equipment designed to control them fails; consequences: fatalities, injuries, environmental and property damage resulting from toxic releases, fires, explosions, and/or runaway reactions

Table G-4. Copy of RBPS Pillar Descriptions for the Personnel Competency Surveys (Table G-4 in the Excel file for Appendix G)

		Table G-4
		Descriptions of the CCPS Risk Based Process Safety (RBPS) Pillars
Pillar		
	Process Safety	Process Safety is the approach that taken to ensure that the assets used in chemical manufacturing, distribution and handling operations are managed and under control to minimize the likelihood of a loss of containment that could lead to a fire, explosion, exposure, or business interruption. Process safety management starts with a sound design, and requires good systems for hazard identification and mitigation, for training, for operating, maintaining, and changing the operations and equipment, and for preparing for and responding in emergencies.
Commit to Process Safety		
1	Leadership and Culture	Senior leadership must show commitment to creating and valuing a process safety culture. The company's corporate and business leadership must demonstrate a visible and ongoing commitment to overseeing and improving process safety performance.
		Leaders at all levels in the company must demonstrate an understanding of the importance of process safety. Leaders establish and communicate process safety performance expectations, including measurable goals, objectives and targets; allocate sufficient resources to meet performance expectations; and promote an observable culture of process safety across the organization. Leaders must promote and develop a process safety culture within their organizations, encouraging openness in raising concerns and identifying opportunities for improvement.
2	Accountability	Process safety accountability must be established at all levels within the company. Process safety is integral to business operations and stakeholder expectations, helping reduce the company's operational risk.
		Process safety roles and responsibilities at all levels across the organization must be clearly-defined and include an expectation to raise, and authority to respond to process safety concerns. Leaders are held accountable for process safety performance. Employees understand the importance of process safety as it applies to their jobs and are responsible for following and contributing to the work activities to achieve improvement in company process safety performance.
Understand Process Hazards and Risk		
3	Knowledge, expertise and training	Programs must be in place to provide process safety knowledge, expertise, tools and training, as appropriate, to employees managing the process safety risks
		Process safety competency requirements are established and executed for leaders managing the operations, engineering and operational personnel managing the process safety risks (including contractors and third-party service providers), commensurate with the activities performed. Employees and contractors are trained on process safety, commensurate with their process safety responsibilities. Process safety experts are provided for continuing education related to emerging process safety tools and techniques.
4	Understanding and prioritization of process safety risks	Programs must be in place to systematically understand process safety risks throughout the organization, prioritize actions and allocate resources.
		Companies must identify and understand the hazards and risks of their operations. Companies must implement systems for documenting and accessing comprehensive and current information on process-related hazards and risks to enable informed decision making at all leadership levels.
Manage Process Safety Risk		
5	Comprehensive process safety management system	A comprehensive process safety management system must be developed and implemented to manage process risk to drive continuous improvement.
		Companies must design systems to manage and mitigate identified process safety risks with adequate safeguards. Management of process safety must take into account: inherently safer approaches; passive controls; engineering controls; operational controls; inspection, maintenance and equipment integrity programs; management of change procedures; and emergency response planning
Learn From Experience		
6	Information sharing	Programs must be developed to actively share relevant process safety knowledge and lessons learned across the organization, including methods for making information available to relevant stakeholders.
		Companies must establish programs that foster two-way flow of information between management, employees, contractors and other stakeholders to share process safety information. These programs must share the results from relevant process safety reviews, inspections, audits, and incident investigations across the company in a timely manner. The programs should promote sharing of process safety concerns across all levels with out fear of retribution.
7	Monitoring and improving performance	System to monitor, report, review and improve process safety performance must be developed and implemented.
		Leaders at all levels of the company, where applicable, must monitor process safety performance as a part of their responsibilities. Using appropriate leading and lagging indicators, routine evaluations of process safety management systems can be used to confirm that the desired results are achieved. These results must be reviewed at planned intervals to determine progress against process safety performance expectations and to take action to improve performance when needed.

**Table G-5. Copy of the References for the Personnel Competency Surveys
(Table G-5 in the Excel file for Appendix G)**

		CCPS Guideline	Chapter
Table G-5			
Specific CCPS References for additional information on the RBPS Pillars			
1) Survey for understanding Process Safety — Who is accountable and where does it apply		CCPS 2007 (RBPS)	Chapter 2, Overview of Risk Based Process Safety
2) Pillar: Commit to Process Safety			
	1 Process safety culture	CCPS 2007 (RBPS)	Chapter 3, Process Safety Culture
	2 Compliance with standards	CCPS 2007 (RBPS)	Chapter 4, Compliance with Standards
		G/L for Implementing Process Safety Management Systems	
	3 Process safety competency	CCPS 2007 (RBPS)	Chapter 5, Process Safety Competency
		CCPS 2011a (Auditing)	Chapter 6, Process Safety Competency
		G/L for Managing Process Safety Risks During Organizational Change	
		G/L for Process Safety in Outsourced Manufacturing Operations	
	4 Workforce involvement	CCPS 2007 (RBPS)	Chapter 6, Workforce Involvement
	5 Stakeholder outreach	CCPS 2007 (RBPS)	Chapter 7, Stakeholder Outreach
		CCPS 2011a (Auditing)	Chapter 8, Stakeholder Outreach
		CCPS 2010 (Metrics)	Chapter 6, Communicating Results Section 6.3, Different Audiences
3) Pillar: Understand Hazards and Risk			
	6 Process knowledge management	CCPS 2007 (RBPS)	Chapter 8, Process Knowledge Management
		CCPS 2011a (Auditing)	Chapter 9, Process Knowledge Management
		G/L for Engineering Design for Process Safety, 2nd Ed.	
		G/L for Fire Protection in Chemical, Petrochemical, and Hydrocarbon Processing Facilities	
		G/L for Process Safety Documentation	
	7 Hazard identification and risk analysis	CCPS 2007 (RBPS)	Chapter 9, Hazard Identification and Risk Analysis
		G/L for Hazard Evaluation Procedures, 3rd Ed. Layer of Protection Analysis	
		G/L for Chemical Process Quantitative Risk Analysis, 2nd Edition	
		G/L for Developing Quantitative Safety Risk Criteria	
		G/L for Analyzing and Managing the Security Vulnerabilities of Fixed Chemical Sites	
		G/L for Chemical Transportation Safety, Security, and Risk Management	
		G/L for Acquisition Evaluation and Post Merger Litigation	

Table G-5 (continued). Copy of the References for the Personnel Competency Surveys (Table G-5 in the Excel file for Appendix G)

			CCPS Guideline	Chapter
			Specific CCPS References for additional information on the RBPS Pillars	
4) Pillar: Manage Risk				
	8	Operating procedures	CCPS 2007 (RBPS)	Chapter 10, Operating Procedures
			G/L for Writing Effective Operating and Maintenance Procedures	
	9	Safe work practices	CCPS 2007 (RBPS)	Chapter 11, Safe Work Practices
	10	Asset integrity and reliability	CCPS 2007 (RBPS)	Chapter 12, Asset integrity and reliability
			G/L for Improving Plant Reliability Through Data Collection and Analysis	
			G/L for Mechanical Integrity Systems	
	11	Contractor management	CCPS 2007 (RBPS)	Chapter 13, Contractor management
	12	Training and performance assurance	CCPS 2007 (RBPS)	Chapter 14, Training and Performance Assurance
			CCPS 2011a (Auditing)	Chapter 15, Training and Performance Assurance
	13	Management of change	CCPS 2007 (RBPS)	Chapter 15, Management of change
			G/L for Management of Change for Process Safety	
	14	Operational readiness	CCPS 2007 (RBPS)	Chapter 16, Operational readiness
			G/L for Performing Effective Pre-Startup Safety Reviews	
	15	Conduct of operations	CCPS 2007 (RBPS)	Chapter 17, Conduct of operations
			Conduct of Operations and Operational Discipline	
	16	Emergency management	CCPS 2007 (RBPS)	Chapter 18, Emergency management
			G/L for Technical Planning for On-Site Emergencies	
			Local Emergency Planning Committee Guidebook: Understanding the EPA RMP Rule	
5) Pillar: Learn from Experience				
	17	Incident investigation	CCPS 2007 (RBPS)	Chapter 19, Incident Investigation
			G/L for Investigating Process Safety Incidents, 2nd Ed.	
	18	Measurement and metrics	CCPS 2007 (RBPS)	Chapter 20, Measurement and Metrics
			CCPS 2015 (SHEQ&S)	Guidelines for Integrating Management Systems to Improve Process Safety Performance (this reference)
			G/L for Process Safety Metrics	
	19	Auditing	CCPS 2007 (RBPS)	Chapter 21, Auditing
			CCPS 2011a (Auditing)	Chapter 2, Conducting PSM Audits / Chapter 22, Auditing
			G/L for Auditing Process Safety Management Systems, 2nd Ed.	
	20	Management review and continuous improvement	CCPS 2007 (RBPS)	Chapter 22, Management Review and Continuous Improvement / Chapter 23, Implementation
			CCPS 2011a (Auditing)	Chapter 23, Management Review and Continuous Improvement
			CCPS 2010 (Metrics)	Chapter 7, Drive Performance Improvements / Section 7.5, Management Reviews
			CCPS 2010 (Metrics)	Chapter 8, Improving Industry Performance / Section 8.1, Benchmarking

**Table G-6. Example of Survey 1.1 for Appendix G
(Survey 1.1 in the Excel file for Appendix G)**

**Table G-7. Example of the Senior Leadership Survey 2.1 for Appendix G
(Survey 2.1 in the Excel file for Appendix G)**

Surveys for the Facility Leadership and Process Unit Leadership are Surveys 2.2 and 2.3, respectively, in the Excel file (see list in Table G-1).

Table G-8. Example of the Senior Leadership Survey 3.1 for Appendix G (Survey 3.1 in the Excel file for Appendix G)

3.1) Pillar: Understand Hazards and Risks 6 - Process knowledge management 7 - Hazard identification and risk management	**Senior Leadership** - Resourcing Response - Focus on Process Safety	Current State What is in place today?	Evidence How is it monitored and documented?	Gap Is there a gap?	Action Item What should be done to eliminate the gap?
The foundation of a risk-based approach which will allow an organization to use this information to allocate limited resources in the most effective manner					
Q01	Has **Senior Leadership** established corporate guidance for managing both corporate and facility-specific process safety technologies?				
Q02	Has **Senior Leadership** provided adequate resources, both competent personnel and systems, for managing both corporate and facility-specific process safety technologies?				
Q03	Has **Senior Leadership** established corporate guidance for identifying facility specific hazards and managing their process safety risks?				
Q04	Has **Senior Leadership** provided adequate resources, both competent personnel and systems, for identifying facility specific hazards and managing their process safety risks?				

Surveys for the Facility Leadership and Process Unit Leadership are Surveys 3.2 and 3.3, respectively, in the Excel file (see list in Table G-1).

Table G-9. Example of the Senior Leadership Survey 4.1 for Appendix G (Survey 4.1 in the Excel file for Appendix G)

4.1) Pillar: Manage Risk 8 - Operating procedures 9 - Safe work practices 10 - Asset integrity and reliability 11 - Contractor management 12 - Training and performance assurance 13 - Management of change 14 - Operational readiness 15 - Conduct of operations 16 - Emergency management	**Senior Leadership** - Resourcing Response - Focus on Process Safety	Current State What is in place today?	Evidence How is it monitored and documented?	Gap Is there a gap?	Action Item What should be done to eliminate the gap?
Effective execution of risk based process safety tasks is based on risk management systems that sustain long-term accident free and profitable operations					
Q01	Has Senior Leadership provided corporate guidance for facilities to develop, implement, and sustain operating procedure program?				
Q02	Has Senior Leadership provided adequate resources for facilities to develop, implement, and sustain operating procedure program?				
Q03	Has Senior Leadership provided corporate guidance for facilities to develop, implement, and sustain safe work practices?				
Q04	Has Senior Leadership provided adequate resources for facilities to develop, implement, and sustain safe work practices?				

Surveys for the Facility Leadership and Process Unit Leadership are Surveys 4.2 and 4.3, respectively, in the Excel file (see list in Table G-1).

**Table G-10. Example of the Senior Leadership Survey 5.1 for Appendix G
(Survey 5.1 in the Excel file for Appendix G)**

Surveys for the Facility Leadership and Process Unit Leadership are Surveys 5.2 and 5.3, respectively, in the Excel file (see list in Table G-1).

REFERENCES

Albrecht, Karl and Lawrence J. Bradford, The Service Advantage: How to Identify and Fulfill Customer Needs, Dow Jones-Irwin, Homewood, Illinois, 1990.

ACS (American Chemical Society), "Creating Safety Cultures in Academic Institutions: A Report of the Safety Culture Task Force of the ACS Committee on Chemical Safety," First Edition, http://www.acs.org /content/acs/en/about/governance/committees/ chemicalsafety.html (accessed 06 December, 2013)

ACC (American Chemistry Council), [ACC 2013a] http://responsiblecare.americanchemistry.com/Business-Value (accessed 18 September 2013)

ACC (American Chemistry Council), [ACC 2013b] Responsible Care® Management System, RC14001®, http://responsiblecare.americanchemistry.com/ (accessed18-Sept-2013).

ACC (American Chemistry Council) Responsible Care © - Performance Metrics, 2013, http://responsiblecare.americanchemistry.com/Responsible-Care-Program-Elements/Performance-Measures-and-Reporting-Guidance (accessed 18-September-2013). [ACC 2013c]

AIChE (American Institute of Chemical Engineers), Code of Ethics, (accessed 28-June-2013) http://www.aiche.org/about/code-ethics

API (American Petroleum Institute), "Process Safety Performance Indicators for the Refining and Petrochemical Industries," RP 754, 1st Edition, April 2010.

Atherton, J. and F. Gil, Incidents That Define Process Safety, CCPS/AIChE and John Wiley & Sons, Inc., Hoboken, NJ, 2008.

Baybutt, P., "The ALARP Principle in Process Safety," Process Safety Progress (PSP), March 2014, Vol. 33, No. 1, pp. 36-40.

Bloch, K., and B. Jung, "The Bhopal Disaster, Understanding the impact of unreliable machinery," Hydrocarbon Processing, June 2012.

Bond, John, "A safety culture with justice: A way to improve safety performance," Institution of Chemical Engineers (IChemE), Loss Prevention Bulletin 196, 2007.

Broadribb, M. P., Boyle, B., and Tanzi, S. J., "Cheddar or Swiss? How Strong are your Barriers? (One Company's Experience with Process Safety Metrics)," 2009 Spring Meeting & 5th Global Congress on Process Safety (GCPS), AIChE, Tampa, FL.

Browning, J. B., *Union Carbide: Disaster at Bhopal*, from Crisis Response: Inside Stories on Managing Under Siege, edited by Jack A. Gottschalk, Visible Ink Press, a division of Gale Research, Detroit, Michigan, 1993.

Caropreso, Frank (ed.) Making Total Quality Happen, Report No. 937, The Conference Board, Inc., New York, N.Y., 1990.

CCPS, "Layer of Protection Analysis: Simplified Process Risk Assessment," Center for Chemical Process Safety/American Institute of Chemical Engineers, John Wiley & Sons, Inc., Hoboken, New Jersey, 2001 [CCPS 2001].

CCPS, "The Business Case for Process Safety," Center for Chemical Process Safety/American Institute of Chemical Engineers, Second Edition, 2006. [CCPS 2006].

CCPS, "Guidelines for Risk Based Process Safety (RBPS)," Center for Chemical Process Safety/American Institute of Chemical Engineers, John Wiley & Sons, Inc., Hoboken, New Jersey, 2007 [CCPS 2007a].

CCPS, "Guidelines for Performing Effective Pre-Startup Safety Reviews," Center for Chemical Process Safety/American Institute of Chemical Engineers, John Wiley & Sons, Inc., Hoboken, New Jersey, 2007 [CCPS 2007b].

CCPS, "Guidelines for the Management of Change for Process Safety," Center for Chemical Process Safety/American Institute of Chemical Engineers, John Wiley & Sons, Inc., Hoboken, New Jersey, 2008 [CCPS 2008].

CCPS, "Guidelines for Hazard Evaluation Procedures," Third Edition, Center for Chemical Process Safety/American Institute of Chemical Engineers, John Wiley & Sons, Inc., Hoboken, New Jersey, 2009 [CCPS 2009a].

CCPS, Process Safety Beacon, Messages for Manufacturing Personnel, June 2009. http://sache.org/beacon/files/2009/06/en/print/2009-06-Beacon.pdf (accessed 13-Oct-2013). [CCPS 2009b].

CCPS, "Guidelines for Process Safety Metrics," Center for Chemical Process Safety/American Institute of Chemical Engineers, John Wiley & Sons, Inc., Hoboken, New Jersey, 2010 [CCPS 2010].

CCPS, "Guidelines for Auditing Process Safety Management Systems," Center for Chemical Process Safety/American Institute of Chemical Engineers, John Wiley & Sons, Inc., Hoboken, New Jersey, 2011 [CCPS 2011a].

CCPS, "Process Safety Leading and Lagging Metrics," Center for Chemical Process Safety/American Institute of Chemical Engineers, Revised: January 2011 [CCPS 2011b].

CCPS, "Conduct of Operations and Operational Discipline," Center for Chemical Process Safety/American Institute of Chemical Engineers, John Wiley & Sons, Inc. Hoboken, NJ, 2011. [CCPS 2011c]

CCPS, "Guidelines for Managing Process Safety Risks During Organizational Change," Center for Chemical Process Safety/American Institute of Chemical Engineers, John Wiley & Sons, Inc., Hoboken, New Jersey, 2013 [CCPS 2013].

CCPS, "Safety Culture: "What Is At Stake," http://www.aiche.org /ccps/topics/elements-process-safety/commitment-process-safety/process-safety-culture/building-safety-culture-tool-kit/what-is-at-stake (accessed 25-Feb-2015), 2015. [CCPS 2015]

Ciavarelli, Anthony P, "Safety Climate and Risk Culture: How Does Your Organization Measure Up?," Human Factors Associates, Inc., 2007.

Dekker, Sidney, *Just Culture, Balancing Safety and Accountability*, Ashgate Publishing Limited, Burlington, VT, 2007.

DuPont Bradley Curve, http://www.dupont.com/products-and-services/consulting-services-process-technologies/operation-risk-management-consulting/uses-and-applications/bradley-curve.html (accessed 10-Dec-2013).

Gunningham, Neil and Darren Sinclair, "Culture Eats Systems for Breakfast: On the Limitations of Management-Based Regulation," The Australian National University, The National Research Centre for OHS Regulation (NRCOHSR), Working Paper 83, November 2011.

High Reliability Organizing (HRO), http://high-reliability.org/ (accessed 07-Dec-2013).

Hopkins, A., "Thinking about Process Safety Indicators," Safety Science, Vol. 47, No. 4, 2009.

HRO, High Reliability Organizing website, http://high-reliability.org/pages/home (accessed 22-Oct-2013).

HSE, The UK Health and Safety Executive, Successful health and safety management, (Second edition), HSG6, HSE Books (1997). Note: from the HSE website, this book is currently being revised, "Managing for Health and Safety," http://www.hse.gov.uk/managing/index.htm (accessed 19-Oct-2013).

HSE, The UK Health and Safety Executive, "Safety Culture: A Review of the Literature," HSL/2002/25, 2002.

HSE, The UK Health and Safety Executive, "Developing process safety indicators: A step-by-step guide for chemical and major hazard industries," HSG 254, 2006.

HSE, The UK Health and Safety Executive, "The Buncefield Incident 11 December 2005: The final report of the Major Incident Investigation Board," Volume 1, HSE Books, www.buncefieldinvestigation.gov.uk, 2008.

HSE, The UK Health and Safety Executive, "Buncefield: Why did it happen?," The Competent Authority Report, Crown, 2011. [HSE 2011a]

HSE, The UK Health and Safety Executive, "Five steps to risk assessment," Leaflet INDG163 (rev3), revised 06/11 (2011). [HSE 2011b]

HSE, The UK Health and Safety Executive, "Health and safety training, A brief guide," Leaflet INDG345 (rev1), published 11/12 (2012).

HSE, The UK Health and Safety Executive, "Leadership for the major hazard industries, Effective health and safety management," Leaflet INDG417 (rev1), published 06/13. [HSE 2013a]

HSE, The UK Health and Safety Executive, "Leading health and safety at work," Leaflet INDG417 (rev1), published 06/13. [HSE 2013b]

IEC/TC56, Dependability, http://tc56.iec.ch/index-tc56.html (accessed 22-Oct-2013)

Institution of Chemical Engineers (IChemE), http://www.icheme.org/~/media/Documents/icheme/About_us/Royal Charter by Laws Code of Professional Conduct and Disciplinary Regulations-20August 2011.pdf (accessed 19-September-2013)

ISO (International Standards Organization), *The integrated use of management system standards, Edition 1*, 2008. [ISO 2008a]

ISO (International Standards Organization), ISO 9001:2008, *Quality management systems – Requirements.* [ISO 2008b] Note: Revision scheduled for issue in 2015.

ISO (International Standards Organization), ISO Quality Management Series, ISO 9004:2009 - Improving quality management system efficiency and effectiveness.

ISO (International Standards Organization), ISO 19011:2011 - Internal and external quality management systems audit guidance.

Juran, Joseph M., Managerial Breakthrough: A New Concept of the Manager's Job, McGraw-Hill Book Co., New York, N.Y., 1964.

Kane, Edward J., "IBM's quality focus on the business process," Quality Progress, April 1986.

Khorsandi, J., "Summary of Various Risk–Mitigating Regulations and Practices applied to Offshore Operations," Deepwater Horizon Study Group 3, Working paper, http://ccrm.berkeley.edu/pdfs_papers/dhsgworkingpapers feb16-2011/summaries-of-variousrisk-mitigatingregulationsandpractices-jk_dhsg-jan2011.pdf (accessed 09-Mar-2014).

Klein, G., Streetlights and Shadows, Searching for the Keys to Adaptive Decision Making, The MIT Press, Cambridge, MA, 2009.

Klein, J. A., "Operational Discipline in the Workplace," Process Safety Progress, Vol. 24 (4), pp. 228-235 (2005).

Klein, J. A., and B. K. Vaughen, "A Revised Program for Operational Discipline," Process Safety Progress, Vol. 27 (1), 2008. pp. 58-65.

Klein, J. A., and B. K. Vaughen, "Implementing an Operational Discipline Program to Improve Plant Process Safety," Chemical Engineering Progress (CEP), June 2011. pp. 48-52.

Klein, J. A., and B. K. Vaughen, *An Introduction to Process Safety: Key Concepts and Practical Applications*, CRCPress, Boca Raton: Taylor & Francis, to be published in 2015.

Kletz, T. A., What Went Wrong? Case Histories of Process Plant Disasters and How They Could Have Been Avoided, Fifth Edition, Elsevier, New York, 2009. pp. 338-341.

Knowles, R. N., The Leadership Dance, Pathways to Extraordinary Organizational Effectiveness. ISBN 0-9721204-0-8. (2002).

Koch, Charles G., *The Science of Success: How Market-Based Management Built the World's Largest Private Company*, John Wiley & Sons, Hoboken, NJ, 2007.

Leveson, N.G., *Engineering a Safer World: Systems Thinking Applied to Safety,* The MIT Press, Cambridge, MA (2011).

Murphy, J. F. and Conner, J., "Black swans, white swans, and 50 shades of grey: Remembering the lessons learned from catastrophic process safety incidents," Proc. Safety Prog., 33: 110–114 (2014). doi: 10.1002/prs.11651.

Murphy, J. F., "The Black Swan: LOPA and Inherent Safety Cannot Prevent All Rare and Catastrophic Incidents, Process Safety Progress (PSP), Volume 30, Issue 3, September 2011, Pages: 202–203.

Murphy, J. F., and J. Conner, "Beware of the black swan: The limitations of risk analysis for predicting the extreme impact of rare process safety incidents," Process Safety Progress (PSP), Volume 31, Issue 4, December 2012, Pages: 330–333.

National Safety Council, 14 Elements of a Successful Safety and Health Program, 1994.

National Society of Professional Engineers (NSPE), http://www.nspe.org/Ethics/CodeofEthics/index.html (accessed 19-September-2013).

OECD (Organisation for Economic Co-operation and Development), Environment Directorate, "Guiding Principles for Chemical Accident Prevention, Preparedness and Response," Series on Chemical Accidents, No. 10. (2003, revision of the first edition published in 1992).

OECD (Organisation for Economic Co-operation and Development), *Guidance on Developing Safety Performance Indicators related to Chemical Accident Prevention, Preparedness and Response*, Guidance for Industry, Series on Chemical Accidents, No. 19, Paris 2008.

OECD (Organisation for Economic Co-operation and Development), Environment Directorate, "Addendum to the OECD Guiding Principles for Chemical Accident Preparedness and Response (2nd ed.), Series on Chemical Accidents, No. 22, ENV/JM/MONO (2011)15, Paris 2011.

OECD (Organisation for Economic Co-operation and Development), Environment, Health and Safety, Chemical Accidents Programme, "Corporate Governance For Process Safety, Guidance For Senior Leaders In High Hazard Industries," June 2012.

Overton, T., "Meeting Today's Societal Expectations: The Use of Process Safety Metrics to Drive Performance Improvements," Responsible Care® Conference, 2008. http://cefic-staging.amaze.com/Documents/ResponsibleCare/RC Conference 2008/ T_OvertonRC_conference_2008.pdf_(accessed 09-Mar-2014).

Royal Academy of Engineering and the Engineering Council, Statement of Ethical Principles, http://www.engc.org.uk/ecukdocuments/internet/ document library/Statement of Ethical Principles.pdf (accessed 19-September-2013).

Scherkenbach, William, The Deming Route to Quality and Productivity: Road Maps and Roadblocks, CeePress Books, George Washington University, Washington, D.C., 1986.

Scholtes, Peter R. and Heero Hacquebord, "Beginning the quality transformation, Part I; and six strategies for beginning the quality transformation, Part II," Quality Progress, July-August 1988.

Sepeda, A. L., "Understanding Process Safety Management," *Chemical Engineering Progress*, August 2010, Vol. 106, No. 6, pp. 26-33.

U.S. Chemical Safety Board (CSB), "DPC Enterprises, L.P., Chlorine Release," Report No. 2002-04-I-MO, May 2003. http://www.csb.gov/assets/1/19/DPC_Report.pdf (accessed 19-September-2013)

U.S. Chemical Safety and Hazard Investigation Board (CSB), Sterigenics, Ontario, California' August 19, 2004, Report No. 2004-11-I-CA, Issue Date: March 2006.

U.S. Chemical Safety and Hazard Investigation Board (CSB), "E.I. DuPont de Nemours & Co Inc., Buffalo, NY, Flammable Vapor Explosion," Investigation Report No. 2011-01-I-NY [US CSB 2011a].

U.S. Chemical Safety Board (CSB), "DuPont Corporation Toxic Chemical Releases," Investigation Report Number 2010-6-I-WV, September 2011. [US CSB 2011b]

Vaughen, B. K. and J. A. Klein, "Improving Operational Discipline to Prevent Loss of Containment Incidents," Process Safety Progress (PSP), September 2011, Vol.30, No.3, pp. 216-220.

Vaughen, B. K., and T. Muschara, "A Case Study: C ombining Incident Investigation Approaches to Identify System-Related Root Causes," Process Safety Progress (PSP), December 2011, Vol. 30, No. 4, pp. 372–376.

Vaughen, B. K. and T. A. Kletz, "Continuing Our Process Safety Management (PSM) Journey," Process Safety Progress (PSP), December 2012, Vol. 31, Issue 4, pp. 337–342.

Willey, R. J., Hendershot, D. C., and Berger, S., The accident in Bhopal: Observations 20 years later. *Process Safety Progress*, 26: 180-184, (2007). doi 10.1002/prs.10191

INDEX